STORM WARNING

The Story of a Killer Tornado

NANCY MATHIS

A Touchstone Book
Published by Simon & Schuster
New York London Toronto Sydney

TOUCHSTONE
A Division of Simon & Schuster, Inc.
1230 Avenue of the Americas
New York, NY 10020

First Touchstone trade paperback edition March 2008

TOUCHSTONE and colophon are registered trademarks of Simon & Schuster, Inc.

For information about special discounts for bulk purchases,
please contact Simon & Schuster Special Sales at
1-800-456-6798 or business@simonandschuster.com.

Designed by Jamie Kerner-Scott

Manufactured in the United States of America

10 9 8 7 6 5 4 3 2 1

The Library of Congress has cataloged the hardcover edition as follows:
Mathis, Nancy.
    Storm warning : the story of a killer tornado / Nancy Mathis.
        p. cm
    Includes bibliographical references.
        1. Tornadoes—Oklahoma—Oklahoma City—History—20th century.
    2. Tornado warning systems—Oklahoma. 3. National disaster warning
    systems—United States. I. Title.

QC955.5.U6M38 2007
363.34'923097663809049—dc22   2006051237

ISBN-13: 978-0-7432-9660-1 (pbk)

For my mother

The close-knit world of the tornado and severe thunderstorm forecaster often seems somewhat demented to those not knowledgeable in this discipline. This apparent derangement is based on our seemingly ghoulish expressions of joy and satisfaction displayed whenever we verify a tornado forecast. This aberration is not vicious; tornadoes in open fields make us happier than damaging storms and count just as much for or against us. We beg your indulgence, but point out the sad truism that we rise and fall by the blessed verification numbers. There is a fantastic feeling of accomplishment when a tornado forecast is successful. We are really nice people but odd.

—The late Col. Robert C. Miller,
U.S. Air Force meteorologist

And, behold, there came a great wind from the wilderness, and smote the four corners of the house, and it fell upon the young men, and they are dead; and I only am escaped alone to tell thee.

—Job, 1:19

CONTENTS

CONTENTS

INTRODUCTION

O N THE FIRST WARM DAY OF each spring, an elderly American Indian woman would grab a hoe and a flashlight and head down into the storm cellar. This six-foot-square concrete bunker doubled as a frost-free refrigerator for the canned beans, peaches, and assorted other fruits and vegetables she had harvested the previous fall. Wielding the hoe like a makeshift guillotine, she cleared the room of any hibernating rattlesnakes, copperheads, or other small creatures that might have found their way into the shelter during the winter.

With the cellar swept clean, the spiderwebs and animal carcasses removed, my grandmother was now ready for the tornadoes and thunderstorms that would surely come in April, May, and June. Well, almost. She had one more weapon in her arsenal. At the first sign of a dark cloud rumbling in from the west, she would take an axe, point it at the clouds, and swing the blade hard into the ground, certain that this bit of native magic would cause the storm cloud to split and keep us safe from the tornadoes. I always suspected there was more to tornado safety than that, but in the 1960s and 1970s, an axe in the ground was just as accurate as the next day's forecast.

I spent many hours as a child in my grandmother's dank cellar, listening to the winds whistle through the cinder-block vents and the hail hammer the tin door, imagining what was happening outside. My grandmother's fear of tornadoes was hardly unique and not at all unwarranted. The small eastern Oklahoma town of Tahlequah, where I grew up, was the capital of the Cherokee Nation. The site was chosen in a valley because it was believed to be protected from tornadoes. That's one of many myths about the twister; in fact, there are no safe locations.

The tornado remains a great puzzle, its many myths steeped in folklore. This book is the life story of one tornado on one day and its consequences—not just any tornado, but the most powerful twister ever to strike a metropolitan area. It is the life story of a tornado researcher and his legacy—not just any researcher, but the most brilliant meteorological detective of the twentieth century. And it is the story of the lives touched with such a harsh hand on May 3, 1999.

Meteorology is one of the most complex of the sciences. Indeed, it took a meteorologist to develop one of the new fundamentals of science: chaos theory. The breakthrough happened in 1961 while American Edward Lorenz was working with a numerical computer model for weather predictions. While attempting to repeat one weather pattern, in order to save time, Lorenz entered only three decimal places, .506, instead of the six, .506127, the computer could store. He entered his sequence of numbers expecting to see the same weather pattern take shape on the screen in front of him. What appeared was a radically different prediction. He'd assumed that the difference of one part in ten thousand would be minimal, that the picture that emerged would be at least similar to what he'd seen before. Instead, the two patterns bore no resemblance to each other.

In 1979, Lorenz wrote a landmark research paper exploring this phenomenon, "Predictability: Does the Flap of a Butterfly's Wings in Brazil Set Off a Tornado in Texas?" Eventually, chaos theory also became known as the Butterfly Effect, referring to a small act that creates great consequences. Like a small break in the clouds over Oklahoma on May 3, 1999. Like a missed target in Japan on August 9, 1945. Like a mother's brief moment of indecision. Like my grandmother burying an axe blade in the ground.

Chaos theory has been much on display since Hurricane Katrina ravaged New Orleans and the Gulf Coast in September 2005. All actions—or lack of actions—have consequences. The hurricane and the tornado are different meteorological animals, but they send us the same message: we are at their mercy, and we ignore them at our own peril.

—NANCY MATHIS

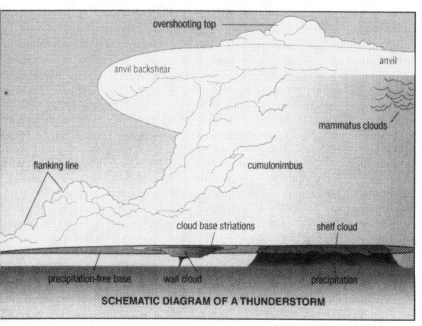

overshooting top

anvil

anvil backshear

mammatus clouds

flanking line

cumulonimbus

cloud base striations

shelf cloud

precipitation-free base wall cloud

precipitation

SCHEMATIC DIAGRAM OF A THUNDERSTORM

(NATIONAL OCEANIC AND ATMOSPHERIC ADMINISTRATION/DEPARTMENT OF COMMERCE)

STORM
WARNING

1

Nature's Atom Bomb

BEFORE THE GREENISH RADAR SCANS, BEFORE blurry photographs from satellites, before television or television meteorologists, and before the snappy twenty-four-hour-a-day Weather Channel, there was this: the faint flicker of lightning and the distant growl of thunder on the prairie's horizon. This was what amounted to a storm warning on the plains.

The far, open sky filled with mountainous cauliflower clouds that grew fat with rain and hail, and those dark olive-hued clouds could conceal the most powerful force known in nature—or not. No one knew for sure, no mere mortal could; after all, the tornado, or cyclone as it was called on the plains, was an act of God, an Old Testament punishment for ill deeds or a test of faith. It was capricious and deadly, leaving the living to bear witness that the great wind came straight from heaven, or so it seemed.

April 9, 1947, sometimes seems like yesterday for Ramona Kolander. Not that she dwells on the day, but it's impossible to forget a day that turns your life upside down. Not even a day really, just a few seconds, and a life's course is altered forever.

"I never would have lived the life I lived without that tornado," she explained. It served as the demarcation in her life, the before and the after.

STORM WARNING

Ramona was a senior at Shattuck High School at the time. Shattuck was a frontier town in far northwestern Oklahoma, situated at the edge of no-man's-land and the Texas border. It was flat and empty, and the emerald fields that bowed with the winds would soon turn golden as harvesttime approached. The Kolanders, like most of their neighbors, grew winter wheat, a difficult, fickle, and occasionally profitable occupation.

For Ramona, most of the day was unremarkable, passing pretty much as it had for the past eighteen years of her life. She remembered the morning as a little cloudy, cool, and damp. She rose at daybreak, dressed, had breakfast, and helped her little sister, LaNita, get ready for school, a monotonous routine that was about to change with her upcoming graduation. She expected she would get a job, probably marry, raise her own kids, and live her life in Shattuck, just as her parents had.

Three Kolander children—Ramona, Floyd, and LaNita—caught the yellow school bus for Shattuck. The baby of the family, four-year-old Doug, stayed at home. The school day became lost to memory, but surely there was some excitement as the class of '47 prepared for its dances, its final tests, and its ceremonies that would send them off into a confident postwar America. The Kolander siblings returned home at 5:00 p.m. Ramona stepped off the school bus and caught a glimpse of a "strange green hue" on the western horizon. She thought little of it. In Oklahoma in the spring, the weather could be unpredictable. The wind blustered harder than usual. A musty odor, a mixture of rain and dust, rode on the air.

The Kolanders lived a few miles outside Shattuck in a white two-story house, a patchwork of adobe and wood frame. There was no electricity. They owned a battery-powered radio but used it sparingly. A giant water tank behind the home provided the family with its drinking water. And, of course, there was a windmill, its sullen metal vanes protesting at every turn with every gust.

There was always a breeze on the prairie. The wind really did come sweeping down the plains as Rogers and Hammerstein wrote in their 1943 musical *Oklahoma!* The inland plains, a great swath of smooth lowlands, etched its way from Canada to North Texas and through much of Oklahoma. The state was a study in topographical transitions. To the east, along the border with Arkansas and the Ozarks, the terrain was hilly, green, and forested. Moving westward, the land rolled gently, and the trees, mainly oaks, grew smaller and sparser. In central Oklahoma, the ground flattened and tilted imperceptibly toward the Rocky Mountains 700 miles farther west. Around Shattuck and far western Oklahoma, the few trees that remained huddled for survival near barely trickling streams, and the tough prairie grass and planted crops took over the leveled landscape. Only windmills, grain elevators, and clouds interrupted the expansive view.

About 140 million years ago, all this land was under the shallow, salty waters of an ocean with a peaceful name—the Sundance Sea. The Rocky Mountains were being heaved upward by smashing tectonic plates, and the Sundance Sea lapped at the mountain range's edge and carried away rich sediments from the Rockies as well as Appalachia and the Ozarks. The ancient sea, which covered the interior of the continent, evenly sifted those deposits like soft flour along its ocean floor as it retreated toward the Arctic. In its wake were the inland plains. The Great Plains encompassed much of the elevated flatlands through the Texas and Oklahoma panhandles. The vast middle of Oklahoma technically was part of the Osage Plains, which, until the pioneers arrived with their plows, was an immense sea of grassland stretching westward. Here, eventually, scientists would figure out how all this strange, Jurassic geography—the arid west of New Mexico, the Rockies in Colorado, and the plains from Texas to the Dakotas—provided an ideal staging ground for the fiercest thunderstorms on the planet.

As America neared the twentieth century, Oklahoma Territory was the last great frontier of unclaimed lands—unclaimed, that is, by white settlers. The federal government began giving away large tracts of the central and western prairie in 1889 in a series of land runs that brought thousands of people to the plains, and whole towns sprang up overnight. Those who jumped the gun, who tried to sneak on the prairie to claim their plots early, were called Sooners. In 1893, the government opened the largest parcel, the seven-million-acre Cherokee Outlet, for settlement. Campsites rose from the grassland. Homesteads, sodbusters, cattle ranches, and railroads all followed. Woodward also sprang to life. Before the land run, Woodward served as a provisioning point for nearby Fort Supply, the U.S. Cavalry outpost that suppressed the last of the Plains Indians. With all the farmers crowding onto the prairie, Woodward began to grow as an agricultural trading post. Thirty miles to the southwest, Shattuck rose from the prairie. It was named for George Shattuck, director of the Santa Fe Railroad, a nod to the railway industry's importance. But the weather should have given the settlers pause. It was unbearably hot in the summer, bone-chillingly cold in the winter, and there was that constant wind.

The Kolanders and most of their neighbors were German. Ramona's grandfather emigrated from Germany to the Texas coast at the turn of the century, where he first tried his hand at rice farming. Soon he followed other Germans to western Oklahoma, where the soil was rich and the land cheap. Broom corn, a variety of sorghum, was the top crop, and Shattuck proclaimed itself the broom corn capital of the United States, back when brooms really were made of dried plant stalks. Her grandfather prospered enough to buy his own land, which he turned over to his son to farm. For the Kolanders, it would mean heartache.

The 1930s were the hottest, driest years on record. The buffalo grasses had held the rich sod to the earth for centuries, but over the decades, the sodbusters' plows upset a delicate balance. When the drought came, there was nothing to hold the soil. There were no crops, only spindly stalks aching for water. The topsoil left by the Sundance Sea simply became airborne. The wind swept the prairie clean, raking away the dirt until all that remained were the heavy, burnt-orange clay sediments.

Bankers, as happened in thousands of cases, seized the Kolander farm at the height of the Great Depression. The Dust Bowl affected the entire southern plains, but no other place came to symbolize the wrathful climate more than Oklahoma. The term *Okies* became the standard pejorative for all migrants who fled the dry plains in search of work and food. But the Kolanders toughed it out, moving to a smaller place and continuing to farm a reluctant land. For days on end, dust clung to the air. On April 14, 1935—Black Sunday—daylight turned to night in the middle of the afternoon. Visibility was zero. The dust boiled across the Texas panhandle and western Oklahoma, so blinding that people caught outdoors needed flashlights to find their way. Total darkness lasted for an hour before the dust eased ever so slightly and faint visibility returned. And those were the storms that Ramona grew up fearing—the great dust storms of the Dust Bowl.

The thunder clapped as the Kolanders finished their dinner. In the family room, Henry Kolander held his two youngest children, LaNita and Doug, on his lap and sang softly to them in German. The children played barbershop and fashioned their father's hair into odd shapes, laughing at the results. Ramona chatted with her brother Floyd, the topic now lost to memory. She remembered that her mother, Stella, must have been doing chores because she rarely sat down after supper.

Lightning grew brighter, and thunder rattled the house. Still, no one was alarmed.

"If God wants it to rain, it will," Doug said as he kissed everyone good night. Ramona led him upstairs to bed, put him in his nightclothes, and began to read him a story.

Woodward was a straight but angled shot from Shattuck to the northeast along the Southern Kansas (Katy) Railroad. With 5,500 residents, it was one of the few towns in Oklahoma that did not lose population during the massive migration created by the Dust Bowl and the Depression. After World War II, Woodward's niche as an agricultural trading hub grew along with postwar America's economic boom.

In Woodward, war veteran Jim Feese prepared for a first date. That morning, he had delivered a refrigerator to a beauty shop and saw a pretty girl. He'd asked her to see a movie that night at the Woodward Theater. *Rage in Heaven* starred Ingrid Bergman and Robert Montgomery. It was a 1941 noir thriller, just making it to Oklahoma, about a psychotic husband who believes his wife is having an affair with his best friend. It was not the best first-date movie, but the only other option was *Hell on Wheels*. On Wednesday nights, the only entertainment options were prayer meetings or the movies. The theater was always packed; so were the churches. The Woodward, at 818 Main Street, was jammed with 300 people. Jim and Reva Valentine found two seats near the back.

The great wind offered no warning. Neither did the government weather bureau. The U.S. Weather Bureau had banned the word *tornado* from its forecasts and warnings a half-century earlier—no need to frighten people. They might have a heart attack and die or, perhaps worse, blame the Weather Bureau for inaccurate predictions.

High above the Texas Panhandle, rivers of opposing air, warm and cool, collided, and ugly gray cloud plumes soared toward the stratosphere. The storm took shape silently, without witnesses, and began to clatter and clang only as it surged upward into the atmosphere. As it took life, the storm began to inhale the prairie winds with such power that its miles-wide underbelly began to rotate, and the heavens opened. Lightning ricocheted across the clouds, and had anyone been watching, they might have seen the wrathful finger of God descending from heaven.

Concealed by heavy rain and large hail, the twister roared from the darkness. It started at White Deer, in the Texas Panhandle, at 5:52 p.m. The tornado was so powerful it blew nineteen boxcars from the Santa Fe Railroad tracks. It next hit the small Texas Panhandle town of Glazier, killing sixteen people, trapped in their homes or businesses, as it chewed the tiny downtown to pieces. A newspaper reported Glazier was "turned to kindling" and "reduced to nothing more than a memory."

The twister moved to Higgins, where it destroyed or damaged nearly every home in the town. Of Higgins's 750 citizens, 51 of them died that night. Most were in a crushed movie theater that collapsed from the winds. The tornado ate the prickly sand sage to its roots, grabbed the barbed wire and telephone lines, and wrapped them into twisted strings strewn along the countryside. Glazier and Higgins became nothing but splinters that littered the barren landscape. Utility poles, railway signals, windmill towers, and trees snapped near ground level.

From Higgins, the tornado, more likely one in a series of twisters, crossed the state line into Oklahoma.

The Kolanders never saw it coming, but they felt it. Ramona was reading to Doug when the house gave a sudden shudder. A second shudder shook the bed. "I think we'd better go downstairs," Ramona told her little brother. She picked him up and headed toward the stairwell.

"The kerosene lamp was on a table by the door and I remembered thinking I should blow this out. I did and at that very moment the floor gave way under my feet and I was enveloped in blackness." Whatever happened next, she did not recall.

The only area weather alarms were sounded by rural telephone supervisors calling each other. A nationwide telephone strike was under way, and many supervisors were filling in for the union employee positions. A telephone operator in Shattuck called a colleague in Woodward: "It's storming out here. Are you-all all right?" Yes, they were fine, replied Grace Nix. An operator from Cestos, southeast of Woodward, called next: "There's a black cloud over Woodward. It looks terrible." Within minutes, the switchboard lights were ablaze with calls, and then nothing. A window in the telephone office burst, and tarpaper, shingles, broken glass, and pieces of an outdoor awning began flying into the room. Nix and assistant Bertha Wiggans took cover under a desk.

At the Woodward Theater, the sound went off midway through the movie. The theater crowd began chanting, clapping, and stomping their feet in protest. Jim took the opportunity to do the big, yawning stretch and put his arm around Reva. As he leaned in for a first kiss, the entire theater went black, and a tremendous roar from the winds outside drowned out the protests and rattled the building.

There was a rush for the exits, but one man stood in the aisle and beseeched the people to stay put. They were safer inside the theater than outside on the street, he argued. The crowd settled down uneasily as the theater creaked and moaned from the wind.

At 8:42 p.m., the tornado, estimated at 1.8 miles wide, entered the western edge of Woodward and churned across its northern neighborhoods. It obliterated 100 blocks of houses, shops, power plants, a factory, a lumberyard. What the tornado didn't destroy, fire did. More than 1,000 homes were damaged or destroyed.

STORM WARNING

The Oklahoma Gas and Electric plant was located on Eighth Street, near downtown. While coworkers ran for shelter, Irwin Walker ran to the control room as the big wind tore through the city's edge. He threw the master switch into the "off" position, thereby cutting power to dangling power lines that otherwise could have sparked more fires and more deaths. Seconds later, the twister leveled the plant. Rescuers later found Walker's body under a pile of rubble.

After Ramona Kolander regained consciousness, she found herself standing in the yard. She was confused and tired, her entire body heavy with fatigue. She sat next to a mulberry tree and leaned her head against the trunk to nap. A constant "horrible moan" interrupted her. It did not stop. The lightning flashes lit her way around the yard, where she found Floyd and LaNita trying to lift an adobe wall off her father. The wall, an interior support from the home, pinned him to the ground from the chest down. It was too heavy for the children to budge.

"Have you seen your mother and little Doug?" he asked. No, she said. "They must be dead," he answered.

She and Floyd ran to the family car. The car and the water tank were still there, but everything else—the house, the barn, the garage—was gone. The splinters, nails, and glass tore at Floyd's bare feet. They fumbled for the car jack in the trunk. She put eight-year-old LaNita into the car's front seat for safekeeping.

The two teens put the jack under the wall and pumped furiously on the handle. They tried to lift it themselves. It was useless. The adobe crumbled at the edges, and the wall still did not budge. Ramona's father told her to go get help. Caked in mud and blood and still in her nightclothes, she ran to the home of the closest neighbor, the Ehrlich family, a mile to the north.

She ran and ran and paused at a bridge over a small stream until a flash of lightning showed her the bridge was still there. The Ehrlichs' home had not been touched. Breathless and feeling faint, she pounded on their door begging for help. And then she collapsed. The Ehrlichs' telephone still worked, and they put out a call to neighbors. The rescue team quickly rallied to the Kolander farm, and the men lifted the wall pinning her father to the ground.

At her father's feet, under the same wall, was her mother. Except for a bruised lip, there were no marks on her body. Her neck had been broken. The men began tossing aside the remains of the Kolander home that were scattered about the yard. They lifted a door and found little Doug crushed to death.

Shattuck Hospital's first indication of catastrophe came from a Higgins farmer who wheeled his pickup to the emergency room door. The back of the truck was filled with wounded Texans. The Shattuck Fire Department and scores of volunteers headed west toward Texas to aid Higgins, unaware of what was happening to the east, unaware that, like a locomotive, the tornado had followed the old Katy railway tracks from Shattuck to Woodward.

When Jim and Reva emerged from the dark theater, downtown was lit brightly by the flames destroying Sharp Lumber Yard and Big 7 Electric Co. He took her home, which had been spared by the twister, checked on his parents, and picked up his brother to join the rescue efforts.

The Stewart family lived on Eleventh Street in Woodward. They were lucky enough to have a basement, and that's where they hid as the twister marched through their neighborhood. Chuck Stewart was five years old at the time. After the tornado destroyed the house above them, his mother and sister boosted him out of a basement window. For a moment he was all alone, just him and the howls in the distance. "There were screams everywhere. We could hear people who had been injured screaming; it was dark and the street was covered with debris," Stewart recalled.

Immediate rescue efforts were ad hoc, neighbor helping neighbor. As the night wore on, Woodward residents formed five-man teams to walk through the hardest-hit neighborhoods calling out for survivors and digging by hand for the injured. The tornado had stripped the trees of their limbs and bark, snapping the trunks about six feet high. Rescuers left the dead in the streets; they couldn't take the time to move them. The people trapped under debris required their immediate attention.

The phone lines were down. The electricity was off. For a while, no one outside of the town knew of their tragedy.

An enterprising telephone lineman, L. L. Avrell, followed the downed telephone wires until he came to lines that were still intact. Avrell shimmied up the telephone pole with his tester phone and managed to reach an Oklahoma City telephone operator with a call for help. The Oklahoma Highway Patrol also radioed for assistance, and ambulances throughout western and central Oklahoma were on their way.

The Daily Oklahoman carried an early-edition front-page bulletin: "The Oklahoma Highway Patrol reported that Woodward, northwest of Oklahoma City with a population of 5,500, was hard hit by a tornado Wednesday night. Jim Holland, Highway Patrol trooper, called his headquarters here from Woodward to say that 'half the town has been blown away and that all possible aid—doctors, nurses and troopers—is needed.' Communications in this northwest section went out after his call, and other details still were lacking at 10:30 P.M."

As word spread, volunteers from nearby towns hopped in their trucks and headed to Woodward to help. Tinker Air Force Base, just outside Oklahoma City, arranged a predawn flight of medical personnel and equipment to Woodward and Shattuck. The Katy Railroad dispatched a train to evacuate the injured from Woodward to hospitals in nearby towns.

The Woodward hospital was overwhelmed. Rescuers left the injured sprawled across the hospital lawn where the doctors and nurses triaged the most critical. When the hospital filled its beds, it moved the injured to the sanctuary at a Baptist church and finally ousted guests at a nearby hotel.

For many World War II vets, the scenes brought to mind the battlefields they had only recently left: the body-crushing, dismembering deaths; the screams of pain; and the pleas for help. The site of the hospital triage was as horrific as anything they had seen in battle. A transport from Tinker Air Force Base ferried some of the injured to Oklahoma City. General Fred Borum, the Tinker commander, surveyed the victims after they landed. During the war, Borum flew wounded Allied soldiers out of France. "I never saw anything worse overseas," he told reporters.

A tornado batters and shreds the human body—not the twister actually, but all the things inside the wind funnel: gravel from the roads, utility poles, furniture, lumber from torn houses, and road signs all become swirling missiles bombarding the body.

Still covered with dirt and blood, Ramona awoke in a bright corridor of Shattuck Hospital, lying on the bare floor. Neighbors had driven her to the hospital, as they had her father. Scores of other bleeding, moaning tornado victims lay beside her, filling the hallway.

"I was in pain; I had several deep cuts and numerous hard blows to my head and body but no broken bones," Ramona recalled. The eldest Kolander sister, Betty, married and living in Texas, was contacted by neighbors and rushed to the hospital trying to find her family. Betty overheard nurses discussing the injured in Ramona's ward. They should be treated last, one nurse told another, because they were unlikely to live.

"When I did not die, they finally cleaned my cuts, stitched them up, and put me on a cot on the floor of the clinic with the other less seriously injured. The next day one of my classmates saw me but did not recognize me because of all the black and blue bruises on my swollen face."

THE DAILY OKLAHOMAN SURVEYED the Woodward destruction—the lots scraped clean, the trees turned to stumps, homes turned to scrap, the number of dead—and in the postwar hyperbole of the time called it the "Sooner Hiroshima." The tornado, it said, was "nature's atom bomb."

The bomb as metaphor was still new. Yet the Oklahoma newspaper was not alone in noting a similarity.

The man destined to become the twentieth century's greatest meteorological detective knew the power of the bomb. It would teach Tetsuya Fujita many things about the nature of winds and the nature of luck. On August 9, 1945, the twenty-four-year-old college professor hurried calmly, along with Meiji College students and colleagues, toward an underground bunker next to the physics building. The air raid siren screamed an alert. A B-29 bomber dubbed *Bockscar* flew overhead, hidden by the clouds, one bomb clutched in its belly.

War defined Fujita's earliest years—not the sounds of air raid sirens that pierced the silence around Meiji College on August 9, 1945, but the government secrecy that blacked out area maps, banned weather statistics, and withheld all manner of scientific data that piqued Fujita's vivid imagination as a child and his quest to know the unknown, to see the invisible. He made his own observations, his own maps, and his own experiments, first with the ocean, then with the sky, and eventually with the weather.

Fujita grew up in Nakasone, a small village of 1,000 people on the island of Kyushu, the most southwestern link in the chain of large islands that makes up Japan. Green, soggy fields of rice surrounded the city, as did two active volcanoes. His father, Tomojiro, was an elementary school teacher, and the family lived comfortably in a home originally built by his grandfather. In kanji, the character "Tetsu" means philosophy and "ya" denotes a boy's name. His father gave him his name, Tetsuya, the philosopher. He took the name seriously.

Nakasone was like any other small town. Visitors entered unlocked homes uninvited, with merely a "Mr. Fujita, I am here, where are you?" Customers charged their purchases at Nakasone stores and paid all debts by December 31 each year. On New Year's Day, people celebrated their birthdays. There was, at least in Fujita's memory, no sense of hardship in this small town so far from Tokyo.

Even as a child, Fujita had a precocious intellect. He was immensely curious, observant, and imaginative. He watched the rising and falling tide near his home, so he could run along an exposed sandbar to catch clams and shako fish. "At that time, I was interested in astronomy because the variation of the sea surface was closely related to the relative position of the sun and the moon." He was no ordinary elementary school student. Confident in his ideas, Fujita was unafraid to offer contrary opinions even when it brought him trouble.

On a school field trip as a teenager, he visited the Yabakei Rapid, where a Buddhist monk, Zenkai, spent thirty years digging, with only a hammer and chisel, a tunnel through a cliff over the rapids. Asked to express his admiration, Fujita suggested the monk should have spent fifteen years building a digging machine so that the tunnel could be completed in another fifteen years, thereby leaving behind both a tunnel and a digging machine that could be used to build even more tunnels. "Unfortunately, I did not receive a passing grade because I failed to appreciate the monk's spiritual accomplishment."

Exploring and cartography were his hobbies. As a teenager, he mapped the contour lines of his school yard. He fashioned a telescope to track the spots on the sun. He and a teenage buddy discovered a limestone cave while climbing a hillside near his home. Carrying only survey chains and a flashlight, he mapped the entire cavern.

He wanted to attend Hiroshima College of High School Teachers, but his ailing father asked him to stay closer to home, to attend Meiji College of Technology. Meiji College specialized in mining, metallurgy, and mechanical engineering. Fujita opted for the last. His engineering emphasis was on the measurement of impact forces.

He also worked part time for a geology professor and was assigned the task of drafting topographic maps of four volcano craters on the island. "After working on his project for several months, my eyes began seeing contour maps as if they were three-dimensional mountains." He considered changing his major to geology, but his father died in 1939 and his mother was ill. He didn't want to be away from her for extended research trips. He also was not one for strenuous hiking or lugging interesting rocks.

His ability to envision three-dimensional forms led to a Japanese Navy contract on ways to detect U.S. aircraft. The mission involved the effect of the curvature of the Earth and weather on multiple searchlights and three-dimensional triangulations. The idea was to determine how the results could be used to locate U.S. warplanes through a new Japanese tool called a radio-locator, Japan's term for radar. The U.S. radar was far more sophisticated and advanced.

Fujita would not have time to finish his project.

STORM WARNING

The war came to Japan's home front in 1945. That spring, Fujita visited an old school friend in Tokyo. His visit coincided with a new American plan to force Japan's surrender. More than 300 bombers ravaged the capital city for two days in March with a new type of bomb. The bombs carried napalm, an incendiary jelly that set sixteen square miles of the city ablaze. Within three hours, nearly 100,000 Tokyo residents died, the stench of their burned flesh settling over the island.

Fujita and his friend ran from the house as they heard the air raid sirens. They glimpsed fiery red skies to their southeast and heard the explosions growing ever closer as the B-29s dropped their bombs at will. "Early the next morning we walked around the neighborhood and found unexploded cylinders of incendiary bombs stuck deep into the gravel road near his house."

He kept his emotions to himself and his mind on science. In his memoirs, he simply noted, "Several days later, while on the train from Tokyo back to Kyushu, I began to think about a mechanical/electrical analog computer operated by converting mechanical quantities into electrical signals." He focused on his radar project. Surely the deaths weighed on Fujita as they did other Japanese. A U.S. invasion was obviously imminent, and Japan's defeat at Okinawa provided the Americans with a staging area. But in its 2,000 years as a military force, Japan had never surrendered. Indeed, there was no word for *surrender* in its language. Until August 1945.

On August 6, the *Enola Gay* dropped the first atomic bomb on Hiroshima as the United States sought Japan's unconditional surrender. An estimated 70,000 to 80,000 people died instantly from the fireball and blast wave. "We have spent two billion dollars on the greatest scientific gamble in history—and won," President Truman told Americans. He again called for Japan to surrender: "If they do not now accept our terms they may expect a rain of ruin from the air, the likes of which has never been seen on this Earth."

Three days later, Fujita, an assistant professor of physics at Meiji College, was hurrying toward a bunker on a cloudy morning. *Bockscar*, carrying a 10,000-pound atomic bomb, could not be seen or heard from the ground. The air raid sirens howled as the bomber passed repeatedly over the city, looking and waiting for a small break in the clouds.

"I clearly remember hearing a series of air raid sirens on the day of the bombing, but the aircraft was not visible due to a thick layer of stratus clouds."

The B-29's initial target was Kokura Arsenal, three miles from Meiji College. The bomb ferried by *Bockscar* was even bigger than the one dropped on Hiroshima.

But the stratus clouds, those thick, heavy blankets of icy particles that block the sunlight and make for a dull, gray day, kept *Bockscar* from locating its primary target. The clouds spared Fujita. The bomber turned toward its secondary target, the port city of Nagasaki 135 miles away. An estimated 35,000 to 40,000 people died as the equivalent of 21,000 tons of TNT dropped onto the city. Japan surrendered unconditionally.

The unexploded Tokyo firebombs, obeying his father's wishes to attend Meiji instead of Hiroshima College, the thick stratus clouds: luck provided Fujita the only shelter he needed.

A month later, he stood amid the devastation that had been Nagasaki and Hiroshima. The Japanese government sent teams of Meiji College engineers and students to study the bomb sites. Charred bodies still littered the hillsides. Radiation dusted the cities, and people were developing mysterious illnesses. Radiation sickness was a surprising by-product of an atomic explosion. "Do not sit on or touch anything in the bomb area," Fujita advised his students. Some of them nevertheless became ill.

At Nagasaki, Fujita used his expertise on impact forces to help determine the exact location of the explosion. He intended to use the same principle for triangulating radar beams to determine the flash point of the explosion, but the massive devastation left him with few objects from which to plot the angles of the shock wave. Walking around a cemetery, Fujita found a bamboo flower pot with a crescent-shaped burn mark on the inside lip. He scurried to other cemeteries surrounding the city. The flower pots all had the same burn marks. He used these flower pot shadows to angle back to ground zero. His theory was that the bomb had exploded 520 meters above Nagasaki. Using the same technique, he estimated Hiroshima's explosion at 530 meters. Both estimates were close. The bombs detonated in the air, not on contact with the ground. He surmised the United States knew the air pressure at each city, which it did, triggering the explosions above the surface.

At both bomb sites, Fujita noticed a giant starburst pattern directly beneath the explosions. The trees at Nagasaki were burned but still standing directly beneath the explosion. Likewise, a bridge that had been *Enola Gay*'s target still stood. The outburst from the bombs flattened bamboo, pine trees, and steel poles in a circular pattern away from ground zero. He envisioned the explosion, the great white blinding ball of fire and an incredible whoosh of winds that fell to the Earth and spread outward in all directions.

This Nagasaki starburst pattern intrigued him, even more so when he saw the pattern repeated years later in the middle of the United States after a thunderstorm.

For Fujita, the worst days were after the war. Inflation soared by more than 200 percent. Rice became so expensive that the college professor could barely afford it. Meiji College even offered "resting days" so that famished staff members could conserve their energy.

"My daily life under the postwar inflation was miserable." He turned to weather not out of a great love of the science, although it became that, but as a secondary income source. In 1946, he received a grant to reeducate grade school teachers in any science topic. He chose weather science because it could be studied cheaply. All he needed was a pencil and paper and his immense powers of observation.

He collected daily weather data—wind speed, temperature, air pressure, and humidity—to prepare a monthly booklet for schoolteachers. In the spring of 1947, Fujita learned silk-screening in order to meet the teachers' demands for color weather maps. His cartography skills dazzled the teachers.

While collecting weather statistics, Fujita became interested in thunderstorms—their power, their lightning, and their mystifying mechanics.

The mere clap of thunder sent him scurrying to the rooftop of his boyhood home, armed with his notebook and pencil. One evening, Fujita recorded the direction of lightning and the time between the flash and the thunder for the next ninety minutes. He charted the location of thirty-three lightning strikes on a map of the island as the storm moved northeasterly toward the seashore, determined their grouping, and suggested that each of three identifiable groups represented the heart of the thunderstorm.

Money was scarce in postwar Japan, and there certainly was none to spare for scientific research. Yet there was plenty of *kagakusuru kokoro*— a spirit of pursuing science. No one had that spirit more than Fujita. He cajoled, recruited, and enticed an army of volunteers from his students, their friends, and their relatives. A solar eclipse over the island presented an opportunity to study the effect of the eclipse on wind and temperature. And Fujita the organizer amassed his first armada of observers, a plan he would use repeatedly in the years to come.

As the solar eclipse approached, Fujita deployed his volunteers to forty-six observation stations and assigned them the task of recording weather observations in fifteen-minute intervals for eight hours. Their only tools were thermometers, barometers, and handmade wind socks. Three eager students volunteered to scale the Japanese Broadcasting Association transmission tower at one-, thirty-, and fifty-meter intervals to track the vertical winds.

The volunteers' observations allowed Fujita to illustrate numerous charts and graphics on the wind and temperature changes. For the volunteers, it was all for fun, perhaps a nice diversion from postwar miseries. Regardless, the scientific recruits donated their own time and effort. It made for a nice scientific project, but Fujita wanted more.

He wanted to feel the winds.

A hot August day in 1947 appeared promising for a storm. He trekked up a steep slope to Seburiyama Mountain's 3,400-foot peak, where Japan had a weather station and the U.S. Air Force had recently installed a radar center. He initially planned to observe the storm in its elements, but lightning persuaded him to seek the safety of the weather office. From the rickety, leaky weather hut, Fujita recorded as many data as possible: wind speed and direction, temperatures, dew points, and air pressure. He later gathered similar information from thirty other weather stations. His only tools were pencil and paper and his own senses. Fujita sketched a series of time-lapse drawings of the thunderstorm—its height, its winds, the rain and lightning.

What Fujita learned from this one trip to the mountaintop was that thunderstorms spewed cool downdrafts off their backsides while sucking in warm updrafts. The years have dimmed the origin of this "cool downdraft" discovery, but a half-century later, this cool downdraft would prove critical to tornado formation.

Fujita's first research paper, "Raiu-no-hana" (Thunder-Nose), elicited little response from Japanese scientists. Months later, Fujita made a presentation to the Fukuoka Weather Service District, from which he had observed the Seburiyama thunderstorm. He received a warmer reception from the field forecasters—and he received a gift that changed his life.

After the session, a weather service employee told him of fishing a copy of "Nonfrontal Thunderstorms" from the garbage dumped by the nearby U.S. Air Force radar base. The research paper was written by a University of Chicago professor named Horace Byers. Would he like to have it? the employee asked.

Fujita—the scientist, the theoretician, the engineer—had only one explanation for the discovery of Byers's report: pure luck.

Fujita borrowed cash from the mother of one of his college students and bought an English-language typewriter. He translated his own "Thunder-Nose" research paper, pecking with one finger at a time, and mailed the paper to Byers, seeking his opinion and comment. In Byers, he sensed a kindred soul, a scientist as interested in the powers and mystery of the thunderstorm. He waited anxiously for a response from the Chicago scientist.

His father always said nothing in this world remains unchanged. "Look up at the full moon in the sky," his father told him. "It will have to turn into a new moon."

For the world of meteorology, a new moon was at hand. One near-sighted Japanese immigrant with a preternatural ability to imagine the invisible was about to turn his formidable skills toward nature's most powerful storms.

SATURDAY, THREE DAYS AFTER the tornado, Woodward began burying the victims. The weather again tormented the dead and the living. The skies grew dark gray, and the rain poured steadily all day, occasionally turning to sleet and snow just to add to the misery. At the cemetery, volunteers clawed through the mud with shovels, working from dusk to dawn in freezing rain to dig enough holes for all the dead.

The two Woodward funeral homes each held a funeral every hour on the hour for the entire day. The schedule repeated daily, until a few of the least damaged churches were repaired and hosted funerals for their parishioners. A constant parade of hearses and ambulances ferried the dead to their final resting place. Tractors had to pull some of the hearses through the muck. The first to be buried were two boys, the only children of Mr. and Mrs. H. C. Harper. Roy Lee was four and his brother H. C., Jr., just two months old. The Salvation Army's state commander presided over the graveside ceremony. One body, a baby girl, was never identified and was buried beneath a headstone that merely listed her as "unknown victim." One four-year-old girl simply disappeared, never to be found or buried.

Officially the toll stood at 181 in three states, with 107 from Woodward. It took weeks to find and count them all. Thirty horsemen rode up and down the banks of the North Canadian River outside Woodward in search of missing people. The Salvation Army and Red Cross kept a casualty list. Much of Woodward was in shambles. Artillery units from Fort Sill, outside Lawton in southwestern Oklahoma, pitched tents for the homeless and set up a field kitchen for the hungry. H. E. Bailey, the state transportation director, brought in equipment to help clear the roads and clean up the tons of trash.

For the state disaster organizations, it was a familiar drill. Tornadoes were a fact of life, though never as devastating as the Woodward tragedy. Two years earlier, a twister had killed sixty-nine people in the small town of Antlers. In '42, two massive tornadoes, occurring two months apart, claimed fifty-two people in the small eastern Oklahoma town of Pryor and thirty-five in Oklahoma City. There were ninety-seven killed in Snyder in 1905 and seventy-one in Peggs in 1920, both tiny communities that lost a substantial percentage of their populations.

But Woodward was the state's worst. A generation of Oklahomans grew up recalling the Woodward twister. People gathered on front porches on humid spring days and told stories about the horror, the deaths, and the survivors. As a boy, Gary England, who would become Oklahoma's premier television weatherman, would be spellbound by these tales: A large truck never found. Clothes ripped off victims by the wind. A naked body wrapped around a utility pole.

He and his family were living near Woodward at the time. He remembers the setting sun tinting the sky on April 9, 1947. The sky appeared so unusual that England's family and neighbors gathered at a nearby school yard to stare in wonder at the clouds on the western horizon. Gary recalled that the skies, from his vantage point, "looked like fuzzy, pink egg cartons floating gently across a darkening sky." To meteorologists, as Gary later learned, these clouds were called mammutus clouds, and they indicated a large thunderstorm nearby.

His father made the first weather forecast Gary remembers hearing: "Somewhere tonight, there's going to be a bad tornado." He was right.

The 1947 tornado ranked sixth among the nation's top ten deadliest twisters.

From 1900 through 1940, these explosive by-products of thunderstorms claimed more than 8,700 American lives. Still, the U.S. Weather Bureau officially remained silent. There would be no tornado forecasts; the bureau wasn't even sure if forecasts were possible.

But 1947 was an unusually deadly year for tornadoes nationwide. According to the U.S. Weather Bureau, forerunner to the National Weather Service, 161 tornadoes across the country killed 306 people, one-third of them from Woodward's lone funnel.

The Woodward twister would be the first of a series of deadly postwar tornadoes to kill 100 or more people. Still, Weather Bureau forecasters, on orders from Washington, would not issue tornado forecasts or warnings. The United States had just won a global war, unlocked the secrets of the atom, and was the major military power in the world—but it would not utter the word *tornado*.

At most, local officials might warn of "severe local storms." No field forecasters dared use the word *tornado* for fear of inciting not just panic but their supervisors. For the most part, the residents of the Great Plains had to rely on their own common sense, their own eyes, and their own luck.

THEY BURIED RAMONA'S MOTHER and brother in the same casket. Her father told her that as the storm grew stronger, he and her mother had leaped from their chairs. "Those kids are upstairs," were her last words. She was ahead of him, rushing toward the staircase, when a wall collapsed on them. The door that crushed Doug broke "nearly every bone in his tiny body."

The Kolanders, like many other rural families, had an underground cellar where they kept canned vegetables and supplies, and they had the radio, though no weather bulletin was issued. The cellar also served as storm shelter. They could have made it to safety; her mother and brother could have been saved, and her life would have been so different.

If only, said Ramona, they had been warned.

2

A Meteorological Star

THE CHASE BEGAN, AS ALWAYS, WITH an argument.

"We should have stayed on Highway 9, and we'd be there sooner. We're zigzagging all over the damn place. We're going to miss the hoses. It'll quit producing by the time we get there," Rich, the passenger, complained.

"You're the navigator," snapped Roger, the driver. "Besides, at least this way, we know we're staying south of the core."

"We could drive all the way to Dallas and stay south of the stupid core. This is what I get for listening to you. We'd have been all right if we stayed on Highway Nine."

"Maybe, but what are you going to do? Rewind the clock? Shut the hell up, Hoss."

Pocked with rust and old hail dents, the aged, creamy-white Pontiac Parisienne station wagon shimmied across the swells of an Oklahoma farm road, its thinning rubber wipers smearing the light rain across the windshield, but clearing just enough to give a glimpse of the dark storm cloud ahead. Until now, it had been a quiet spring, and this day, Monday, May 3, 1999, was the first decent chance to chase storms. Inside the Meatwagon, the 1986 Parisienne's nickname, the AM radio simulcast an Oklahoma City television station's storm report. A wispy tornado was digging into a cattle pasture miles ahead of them to the west, and Rich and Roger were missing it. They could hear the excitement in the storm tracker's news report on the radio: "There's a small one on the ground. Whoa, there goes something! It's beautiful."

Roger Edwards gripped the Meatwagon's skinny steering wheel with both hands, cautious and alert. Rich Thompson had the map, "The Roads of Oklahoma," spread across his lap. But Roger had the wheel, and he intended to give the storm cloud ahead a wide berth, just to be on the safe side. It already was barfing out tiny twisters, and Roger didn't want to drive too close without an escape route in case the tornado caught them unaware. Hail dings in the heavy-gauge sheathing of the Meatwagon served as reminders of past navigation errors.

Not that a few new pings would have mattered to the Meatwagon. It moved with the muscle of an artillery tank accessorized by a topside luggage rack. Its faux wood side paneling had long ago flaked off in chunks, the driver's-side window didn't work, the passenger-side door had to be kicked open from the inside, and whatever it was that manufacturers stuff into the interior roof lining sprinkled out with every bump. It was the perfect "storm-intercept slash fishing vehicle," as Roger called it.

Like the Meatwagon, Roger was stoutly built, but without the wear and tear of 180,000 miles on the odometer. Still boyish at thirty-one, he had thick brown hair that turned curly at the ends when it grew too long and soft brown eyes that warmed to any scientific topic, but especially to tornadoes. Rich was smaller, more lithe, like a runner. He was more of a gabber, but with the same love of weather in general and twisters in particular. Both were Texas natives—Roger from Dallas and Rich from Houston.

In the weary Parisienne, they looked like two guys down on their luck, but they chased storms and hurled insults only in their spare time. At work, they coauthored reports with engaging titles, such as "Nationwide Comparisons of Hail Size with WSR-88D Vertically Integrated Liquid Water and Derived Thermodynamic Sounding Data." Roger and Rich were two of the nation's top storm sentries. They were meteorologists and forecasters at the Storm Prediction Center (SPC) based in Norman, Oklahoma. The SPC was the nation's early warning system when it came to severe weather. Whether it was blizzards in the Rockies, floods in the Midwest, droughts in the West, or storms on the Great Plains, their job, the job of the SPC, was to provide the advance heads-up to the local offices of the National Weather Service. Whatever the atmosphere could conjure—except hurricanes—was within the province of the SPC. Its better-known sister agency, the National Hurricane Center, based in Miami, Florida, handled the hurricanes. All were under the bureaucratic umbrella of the National Oceanic and Atmospheric Administration (NOAA).

The meteorologists at the SPC were the most skilled and experienced of all the forecasters at understanding and detecting the probability of severe storms. And *probability* was still an operative word. Forecasting, even with SPC's satellites, radars, and supercomputers, was at times as much art as science, as much intuition, common sense, and experience—or fuzzy logic as it's called in physics—as cold, hard data, especially when the data conflict.

Since 1996, when it moved from Kansas City, Missouri, to Norman, SPC operated from a cramped, brightly lit first-floor office in a nondescript building at Max Westheimer Field, a small airport that had been a naval aviation training center during World War II. The SPC was shoehorned into the same building with the National Severe Storms Laboratory, the federal government's storm research agency. Once inside, SPC forecasters were glued to their computer monitors. There was nothing else to look at. A long observation window in a hallway allowed other scientists and visitors to peek at SPC operations, but there was no window to the outdoors. An SPC forecaster who wanted to know what was happening thirty yards away had to look it up on the computer.

The two-story building appeared built on the cheap. It was made entirely of concrete, its interior walls and floors poured from wooden forms. The plywood knots left their own fossilized impressions. An exterior brick veneer added a touch of class. The building's formal address was 1313 Halley Circle, a little weather insiders' joke. Edmond Halley, famous for discovering Halley's Comet, also expounded on the convergence of air that made up the trade winds. All things weather swirled around this little hub.

Across the street, a smaller, almost identical building housed the Norman office of the National Weather Service (NWS). The local NWS office, one of 121 weather service offices nationwide, was responsible for weather forecasts for western Oklahoma and a sliver of North Texas. The only clues that something special might be happening there on Halley Circle were the massive satellites and the Doppler radars, which were sheathed in white plastic and placed atop scaffolds, like teed-up golf balls.

Norman is twenty miles due south of Oklahoma City off Interstate 35, a small, quiet town and home of the University of Oklahoma. The university produced three things very well: national championship football teams, petroleum engineers, and research meteorologists, roughly in that order. All three were necessities: the first for pride and money, the other two for survival. Oil production had long lubricated the state's economy, and weather had always been its adversary. Slowly, over thirty years, there was a convergence of the federal government's weather operations: the Storm Prediction Center, the National Severe Storm Laboratory, the nation's weather radar operations, and the local NWS office across the street, plus numerous joint projects with the University of Oklahoma's School of Meteorology.

All of this meant Norman served as the unofficial weather weenie capital of the world. A weather weenie is someone, as they say in Oklahoma, who is "eaten up with it," someone who cannot get enough of extreme weather, obsessed by storms in all their permutations, and especially severe thunderstorms. One does not bestow the honor of weather weenie on oneself. Membership in this tightly knit club comes only after displays, on numerous occasions, of respect, knowledge, loads of common sense, and the ability to chase a tornado without rear-ending a fellow twister hunter.

Roger and Rich were weather weenies before weather weenies were cool. Severe weather was not only their expertise but their passion. Roger called it "feasting at the smorgasbord of atmospheric violence." Roger felt about severe weather, especially tornadoes, the way some people feel about sports. All manner of statistics measured the game of baseball, and so it was with tornadoes. Wind speed, hail size, shape of tornado (cone, stovepipe, rope, or wedge), time of day, temperature, dew point, pressure, high precipitation, low precipitation. Roger's brain was filled with a tornadic scorecard.

He does not remember a time when he did not have this obsession. As a baby, he crawled to the screen door of his East Dallas home, smiling as the thunder and lightning rattled the house. At age three, he told his mother he wanted to see a hurricane. At age nine, he saw his first tornado, a distant twister he glimpsed between houses as it loped through northern Dallas. Ignoring his mother's shouts to come inside, Roger ran down the street to keep it in view. It was his first tornado chase.

As a teenager, he walked to the downtown Dallas library to check out the latest edition of *Weatherwise* and other weather-related magazines. He didn't quite understand them at the time, but he read them anyway. He checked out Snowden D. Flora's cult classic, *Tornadoes of the United States*, more than thirty times. He did so many school science projects on tornadoes that his teachers finally forbade it, so he did one on hurricanes.

He named his two children Donna Camille and David Andrew, after four memorable hurricanes. If only tornadoes had names. Roger had even met his wife, Elke, at a greasy-spoon diner where a bunch of chasers gathered after a West Texas storm. In the back of the Meatwagon's dusty cargo space was a piece of splintered two-by-four, a souvenir from a recent trip to the Wind Science and Engineering Center at Texas Tech University. Engineers at the center mimic the effects of small tornadoes slamming into houses and end up with the same results: torn, broken, splintered pieces of lumber. Roger kept his memento in the Meatwagon as a boy might keep a favorite keepsake in a shoebox.

It was hard to top Roger in weather weeniedom. But Rich had the same encyclopedic memory for storms and the same obsession with storm chasing. Rich's wife, Daphne, also was a meteorologist, working at the local NWS office. Their son, Nathan, lacked a famous hurricane name but could spend hours watching tornado clips and the Weather Channel. Weather weenies produced kinderweenies.

A Meteorological Star

There was, said Roger, a "stark majesty" to a tornado—raw and powerful and beautiful in its own way. Each differed from the one before. No two looked the same. And for a scientist's mind, there was the puzzle: Just what made these things tick? At a safe distance, it appeared romantic—like an untamed mustang, free and wild. It also was the most dangerous item on the atmospheric smorgasbord. Together Roger and Rich probably had seen more than one hundred tornadoes. Some chasers kept a careful count. Some even placed little twister decals on their car. Roger and Rich found such belt notching annoying and didn't keep count. Each one seemed a special gift, each one potentially providing a glimpse of its secrets. Both men also were expert photographers, always carting around their cameras, still and video, to capture fleeting sightings.

Roger had experienced the worst of severe weather firsthand. He did a stint at the National Hurricane Center, at a time that coincided with 1992's Hurricane Andrew, the most devastating hurricane in history until 2005's Katrina came along. Andrew's northern wall of wind and rain had clipped Roger's Miami apartment. And he worked at the SPC when much of it was still based in Kansas City, just in time for the Great Flood of 1993, when much of the nation's entire midsection was under the storm-filled waters of the Missouri and Mississippi rivers.

Storms made Roger cautious, all of which further explained his overly circuitous route and Rich's consternation. Despite the trash talk, they remained each other's favorite storm chase partner and chased storms just for the pure pleasure. It was something they had done hundreds of times since their days as roommates at the University of Oklahoma. But even they could not conjure up the ferocious display they were about to witness.

IT IS EASIER TO send a space probe to Jupiter than it is to forecast the weather. Some of the world's fastest computers—those processing trillions of pieces of data per second—are devoted to meteorology. Chaos theory—the idea that the most infinitesimal change can result in wildly varied outcomes—comes from meteorology. Meteorology is full of predictions and probabilities and very little certainty. It's all math and physics and fluid dynamics, thermal dynamics, and various other dynamics that make it one of the most difficult sciences to master. And, in the end, it sometimes comes down to a hunch, to experience, to an educated guess.

The safest educated guess a meteorologist can make is that at some point during the year, there will be a tornado in Oklahoma and the plains. Nowhere else but America's midsection do the atmosphere and geography partner for such a combustible environment. Nowhere else does a large north-south mountain range such as the Rockies loom so close, geographically speaking, to a large, warm body of water such as the Gulf of Mexico. The springtime jet streams also dip down into the Arizona and New Mexico deserts and curve around the southern flank of the Rockies before turning northward, adding to the atmospheric instability.

In Oklahoma, sometime in March or April, the ever-present prairie wind does a 180-degree turn, swinging from the north to the south, hauling buckets of warm moist air from the Gulf of Mexico. The wind shift makes for pleasant mornings and muggy afternoons, perfect for an early golf game, midday gardening, or afternoon storm chasing.

A tornado is the offspring of a thunderstorm, and no other place creates thunderstorms better than the plains. As the warm, moist air pushes north, drier air from atop the Rockies comes screaming toward the plains, itching for a rumble. As every schoolchild learns, warm air rises and cool air falls. If it were that simple, anyone could predict the weather, but scores of complicating factors stand in the way. The atmosphere seeks a constant blue-sky equilibrium and purges itself of anything that upsets its delicate balance. And something is always working against it. The sun creates temperature differences; dips in the air pressure create the winds; a dab too much moisture, and clouds appear. The atmosphere is a study in harmonious chaos. When all the disorderly and mostly invisible ingredients—the wind currents, the moisture, heat thermals bursting from the ground skyward, the tussle between warmer and cooler air—converge in just the right amounts, at just the right moment, in just the right place, the atmosphere purges itself methodically and violently in the form of a thunderstorm.

There are about 100,000 thunderstorms in the United States each year. The collision of rising warm air and falling cool air remains the core of each. Most thunderstorms bring only welcome rain, maybe a little lightning and thunder. Or they join together to form great squall lines and march from west to east across the country. About 10,000 of those thunderstorms are classified as severe, which the NWS defines as possibly producing winds of at least fifty knots, hail of at least three-quarters of an inch, and maybe a tornado. But most storms are of little consequence, spewing rain for half an hour or so before dissolving. They are short-lived and harmless.

On the Great Plains, their beauty, power, and complexity are heightened. Out on the open range, there is an invisible north-south curtain of air that separates the Gulf's warm, moist currents from the Rockies' cooler, dry currents. Called the dryline, the boundary acts like a pre-fight referee. The dryline, mostly peculiar to the arid western plains, is warmer and drier than the air on either side of it, making it fairly easy to locate by looking at the region's temperatures, dew points, and wind directions. The winds on either side of it move in opposite directions. It is along this dryline that the most powerful Great Plains storms form. During the day in the spring, the dryline is shoved eastward. It is felt, not seen. As it passes, the humidity drops, the skies clear, and the winds shift again. Wind aloft pushes the top of the dryline faster than its bottom, causing it to blanket the top of the warm, moist air mass. It forms a cap over the top of the warm, moist air. This is where the real atmospheric battle starts, about 1,000 feet from the surface.

The upper levels of the dryline act like the lid on a teakettle, topping the warm, moist air until the air mass becomes so warm and so humid that it no longer can be held back. It continually jabs until the dryline weakens. This is what the chasers come to see. The warm, moist current bursts through the dryline cap and blasts skyward at 100 miles per hour into the colder dry air above.

The fluffy cumulus clouds form first as the moisture in the warm air updraft hits the cooler dry winds and condenses. The condensation releases latent heat, which creates energy, which creates more vertical speed, which sucks in more warm, moist air, which forces the cooler air to sink ever faster, and the cycle begins. The cumuli begin to fatten and darken. The warm air updraft builds on a column of condensing particles, and the tower of clouds bubbles 30,000 to 40,000 feet in a matter of minutes. The cumulus has become the cumulonimbus, the thunderstorm cloud—Cb in meteorology shorthand.

Because of the dryline storm explosions, the Great Plains also has a propensity to produce a rare and fierce type of thunderstorm: a supercell. It is super in every way. The tower soars to 50,000 or 60,000 feet and bumps against the stratosphere, flattening at the top to give it the distinctive appearance of a blacksmith's anvil. A supercell draws in warm air with such power that the updraft begins to swirl horizontally. The entire back side of the supercell rotates like a 1950s version of a flying saucer. The horizontally rotating mass, called a mesocyclone, is the first clue to radar observers that a thunderstorm could produce a tornado. The mesocyclone produces a "hook echo" on radar, the appearance of a fishhook or a backward letter J as it swirls the winds. The hook echo almost always forms on the storm's rear right side. Chasers call this storm region the bear's cage, where the wind and rain can obscure a tornado. To get caught inside the bear's cage is to risk being eaten alive.

Fully matured, the supercell is more than twice the height of Mount Everest. The lightning begins to ripple—anvil crawlers, they're called—and the thunder echoes. It's the supercell that produces the biggest, most powerful tornadoes.

TORNADOES ARE A PECULIARLY American phenomenon. They can happen anywhere in the world, but 80 percent—roughly 800 to 1,000 a year—occur in the United States. Most occur in the spring, along the stretch of plains extending from Texas to the Dakotas: Tornado Alley. But the first tornado ever recorded in the United States was in the most unlikely of places—Massachusetts in 1671. It wasn't until settlers began moving westward that the tornado became a more frequent and deadly menace. The tornadoes had always been there; there were just no witnesses who reported the events.

Essentially a tornado is merely a vertical column of rotating air, extending from the thunderstorm cloud to the ground. An average-size tornado can produce enough energy to supply an American home with electric power for a year. A hurricane produces more energy because of its size, but pound for pound, the more compact tornado packs a far greater punch than a hurricane. A hurricane can cause more structural damage because its sustained winds last for hours, compared to the seconds of a tornado. A hurricane gives fair warning days in advance; a tornado is completely random, its warning time measured in minutes.

Scientists have some theories but still do not know exactly how or why tornadoes form. They do not know why one thunderstorm produces a tornado and another does not. They do not know why one supercell produces a twister and another does not. They do not know what forces make one tornado more powerful than another or why one twister small in size is just as forceful as a larger one.

A tornado is meteorology's last great puzzle, and its grandest celebrity.

Blame it on Hollywood. The 1996 movie *Twister,* loosely based on the experiments by the University of Oklahoma and the National Severe Storms Laboratory, featured storm scientists gallivanting around Oklahoma in pickups, coming so close to twisters they could practically touch them. The movie helped create a new generation of storm chasers who crowded the back roads and created their own traffic jams as they tailed a tornado. The real veterans—the chasers who predated *Twister* and the IMAX specials and the PBS *NOVA* series that made the hobby so popular—moaned about newcomers' taking the best off-road parking places or creating traffic jams along the two-lane blacktops. Sometimes there were so many chasers it looked like rush hour.

Twister made it look easy, as though tornadoes aplenty plopped from the heavens and one could chase, shower, eat dinner, and go chase some more. But the tornado always has been an elusive creature that regularly humbles the greatest researchers, scientists, and forecasters. That is part of the allure. By May 3, 1999, the twister was no longer just a lone rogue wreaking havoc on the Great Plains. It had its own fan base, its own "torn porn" videos, photographic stills, and T-shirts. Scores of Web sites were devoted to it. Like a media-savvy celebrity, when it made its daylight public appearances, it was followed by a posse of camera-toting aficionados, like paparazzi after a movie star. For the money-to-burn amateurs, there were even guided tornado tours across the plains states during the spring. A ten-day trek cost $2,000 and came with no guarantees.

The whole chase scene dismayed Kansas-born storm scientist Erik Rasmussen: "When I first started chasing I'd be the only chaser on a beautiful storm and just watch it for hours. That was great fun, sitting out in the midst of all the birds singing on the prairie. But then it became this circus of people trying to notch their belts with tornadoes, which just irritates me."

After the movie *Twister* was such a hit, the media focused on the work of scientists like Rasmussen, one of the premier tornado researchers. Tall and skinny, he was dubbed the Dryline Kid by the press for his uncanny ability to find a tornado-spewing storm. The movie, the newspaper stories, and the television specials about chasers and scientists made people excited, and the excited people began showing up to watch the storms. Rasmussen realized it had gone too far when he and a research team chased a storm into the western part of South Dakota's Badlands. The only way to follow the thunderhead was to pay a ten-dollar fee to enter a national park. As Rasmussen and his team drove deeper into the Badlands, they found thirty to forty chasers already ahead of them. "This was truly," he said, "in the middle of nowhere."

He didn't begrudge the storm lovers their chance to see a magnificent thunderhead or even a tornado. But there were just too many of them. Storm chasing, at least for Rasmussen, was no longer fun.

Rasmussen's job, his mission in life, was to discover the Holy Grail of tornado research: tornadogenesis, the mechanics of tornado creation. The question had stumped the greatest minds since Aristotle. And Rasmussen felt he and the team of researchers were tantalizingly close to finding an answer. In the mid-1990s, he was a field general for Project VORTEX, the Verification of the Origin of Rotation in Tornadoes Experiment. It was the largest armada ever assembled for storm research. More than seventy-five scientists, twenty cars outfitted with weather devices, three airplanes, and a hundred volunteers, radar operators, and chasers helped make it the largest U.S. effort to try to unravel the mechanism of the supercell and the tornado.

Unfortunately, the tornadoes refused to cooperate. Its maiden year, 1994, nearly set a record for the fewest number of tornadoes in Tornado Alley; the second year proved more fruitful, tornado-wise. VORTEX captured data on ten tornadoes, including a powerful twister near Dimmitt, Texas, that has become one of the most-studied storms in history.

Even the best chaser will fail to see a tornado 80 percent of the time. That's the frustrating nature of tornado research. It requires not just a scientific mind but infinite patience. And the supercells' big bruisers, the ones that pack winds in excess of 200 miles per hour and live the longest, are the most difficult to catch. The odds of any given square mile in the United States feeling the wrath of the behemoths is more than 12.5 million to 1.

VORTEX-99 was a small offshoot of the two-year tornado intercept campaign. Rasmussen was content to stay at his home base near Boulder, Colorado, and ramrod VORTEX-99 via cell phone and real-time radar. This scaled-down project had access to three sedans outfitted with anemometers (three-cup gauges that measure wind speed and direction), pressure gauges, and thermometers strapped to the roofs, and twelve volunteers, mostly Oklahoma University meteorology students and off-duty forecasters, equipped with more gauges and cameras. There was enough money left in the VORTEX bank account to fund only a few days of chasing. The gamble was about to pay off.

THE BEST ODDS OF seeing a springtime tornado are around Oklahoma City and its southern suburbs. The region may be the most tornado-prone stretch of real estate in the world. One reason is Oklahoma City's size: 622 square miles, once America's largest city in terms of territory. The other reason is the simple law of real estate: location. It is centrally located in a state that is centrally located in Tornado Alley.

No other spot has just the right combination of weather forces and geography, just the right proximity between the cool winds of the Rockies, the warm air from the Gulf of Mexico, and the dry jet streaks (a fast current within the jet stream) coming off the southwestern deserts. In any given year, in any given May, there is a 10 percent chance of having a significant tornado just a little south of the city.

The local NWS counted ninety-nine twisters within Oklahoma City and its close, southern suburbs such as Moore, Del City, and Midwest City between January 1893 and December 1998, using its current city limits as the boundary. And that's not counting smaller tornadoes that no one bothered to record. At least seven violent tornadoes with winds ranging from 200 to 260 miles per hour had trashed the metro area over the years, but fortune had been on its side. The deadliest tornado on record killed thirty-five people and destroyed seventy homes in 1942. At least eighteen times, two or more tornadoes struck the metro area on the same day. The record was five twisters on June 8, 1974.

The first recorded tornado occurred in 1893, just a few years after Oklahoma City's first settlers arrived with the '89 land run. More than 10,000 people had pitched their tents along the banks of the North Canadian River at the Oklahoma Station, which became the city. The first twister killed thirty-one settlers.

Oklahoma City became a crossroads for the railroads and later for the nation's interstate highway system. Interstate 40 served as the nation's major east-west artery, and Interstate 35 linked the Midwest as the north-south highway. Interstate 44, the H. E. Bailey Turnpike, named for the late transportation director, cut a diagonal shortcut between Missouri and Texas. All met at Oklahoma City.

In 1927, oil drillers discovered the vast pool of petroleum beneath the city, and Oklahoma City became an oil town, filled with independent wildcatters who took their own risks and either went broke or became incredibly rich. Oil derricks dominated the skyline and sprang up in residential backyards, all along the highways, and even in grocery store parking lots. A familiar boom-and-bust cycle, so common in farming, continued with the oil industry. The big bust happened in the 1980s. Even the oil well under the state capitol went dry. Banks collapsed and foreclosures soared, just as they did in the Great Depression.

The city was regaining its economic footing when the nation's worst act by domestic terrorists occurred: the April 19, 1995, bombing of the Alfred P. Murrah Federal Building. Not even on the most horrendous day was the state free from the threat of twisters. As the city recovered the 168 dead, tornadoes rampaged through the southern part of the state. High winds from the storms stirred the debris from the Murrah explosion, making it even more dangerous for workers trying to recover bodies.

City leaders tried to revive downtown, an increasingly empty core. The city created an entertainment district from its own brick warehouse section and built a new ballpark for its AAA baseball team, the Red-Hawks. During this time, the metro area enjoyed its longest respite from twisters—nearly five years without a tornado. Perhaps even the tornado gods felt the need to give the metro area a break. But in October 1998, the twisters returned with a vengeance; a rare fall outbreak produced more than twenty tornadoes in the state and several around the city, a national record for that month.

The state itself averaged fifty tornadoes a year, second only to the 120 annually for Texas. But Oklahoma had more significant tornadoes—those with winds above 113 miles per hour—than anyplace else, an average of more than two a year.

Gary England had been watching over the weather from Oklahoma City's KWTV News 9 since 1972. He had issued hundreds of tornado warnings to viewers during his career, probably more than any other TV meteorologist in the country. He even traveled the state with his "Those Terrible Twisters" presentation, filling high school gymnasiums with thousands of people to explain how horrible tornadoes can be and what people, especially children, can do to protect themselves.

STORM WARNING

The weatherman, who as a boy was fascinated by tales of the Woodward twister, told anyone who listened: one day—it might be in a month or next year or in ten years—but one day a really terrible tornado would visit itself upon the metro area. The metro region could not dodge the bullet forever.

KARA WIESE DIDN'T KNOW a dryline from a clothesline. Nor did she care. Kara Wiese had other things on her mind. There would be no weekly T-ball game tonight. She didn't give a second thought about tornadoes or thunderstorms and never had, even though she grew up in Oklahoma with twisters snaking all around her. All she gleaned from the weather report was that Jordan's Monday night T-ball game would be a washout, which meant a restless six-year-old was sitting in the car beside her with nothing to do.

Kara was the team's bench mom—the one who attended every game and sat with the team cheering on every hitter, every score, consoling them at every out. Kara had been a bench mom for most of her life. At age ten, she had started her own neighborhood babysitting service, all under her mother Mary's watchful eye. As a teenager, her resourcefulness as a babysitter earned her a feature story in the local newspaper. She had tended to a toddler with severe asthma who couldn't play outside in cold weather. On a winter day after a heavy snowstorm, Kara carried buckets of snow to the bathtub so the little boy could make a snowman. She had an easy way with children, and kids gravitated to her, a kindred soul.

Now she was twenty-six, raising her own little boy. Jordan was tall for his age, all skinny arms and thin legs, his hair a snowy blond. He was the one man in her life, other than her brother, Dustin, she could count on. Her first love had left her with Jordan. Her second had left her with a pile of bills, a six-figure debt that forced her into bankruptcy after her divorce. All she had in the world amounted to $36,000, and that was a three-year-old black Monte Carlo, five acres of land, a mobile home, and her collection of unicorns.

Kara was a younger version of her mother, barely five foot two, with squinty blue eyes that seemed to disappear when she smiled. Her blond hair fell to her shoulders and her puffy bangs made her look younger than twenty-six.

She lived at Bridge Creek, about twenty miles southwest of Oklahoma City, an easy commute on Interstate 44. At Bridge Creek, there were small, tidy frame homes and a few dozen mobile homes on large lots. Her mobile home addition was called Southern Hills, just pastureland divided into two-acre or five-acre plots. It was a bedroom community for 1,500 folks with decent but not great-paying jobs. They were drawn to the easy commute, the affordability, and the idea of country living so close to the city. There was no mayor or city council; the area was unincorporated.

The only semblance of authority and government was the Bridge Creek School District, which had served as the center of community activity since it opened in 1919. Women used to hold quilting parties at the school, at least until a tornado hit it one evening in the 1950s, damaging the school and injuring a couple of quilters. The Bridge Creek School and Ridgecrest Baptist Church were the main landmarks, as in, "Go down the road to the church and turn right."

Here, the pastureland was pinched with ridges and deep ravines, all gently rolling toward the banks of the South Canadian River. There were two branches of the Canadian River: the north, which ran through downtown Oklahoma City, and the south, which ran southwest of the city. The bramble-filled gullies, which turned into reddish muddy streams during rainstorms, provided Bridge Creek with its name. According to the school district's history, the name Bridge Creek stuck after some settlers cut fresh timber for a bridge across a creek.

Kara's trailer, white with burgundy trim, sat atop a ridge, giving her a view of Bridge Creek proper, if there was such a thing. Visitors had to take a ninety-degree curve on an unpaved road to get to Kara's driveway. She planted big elephant ear plants to hide the trailer's metal undercarriage and lined her driveway with crape myrtles and apple trees that had been birthday gifts from her mother. Violets lined the flower beds around her porch. She had her own landscape vision; she wanted visitors to come around that big curve and be greeted with a forest of floral colors. She wanted that mobile home and her five little acres as pretty as possible.

It was unusually humid when she picked up Jordan at the TLC Day Care Center just before 5:00 p.m. The sun was shining, and Jordan raced to get a small football; they began playing catch in the front yard. The leaves on the trees were a bright green, shiny with the new spring growth. The floral buds from the crape myrtles and apple trees would come later; the trees were late bloomers.

GARY ENGLAND'S WEATHER REPORTS were being simulcast on the radio, and as Roger and Rich listened to Gary and his team of storm trackers chasing the storm, they debated whether they should turn around and head for home. The storm had been producing one small twister after another. Rich thought the storm could turn HP (high precipitation), ruining the chance of seeing a tornado. Heavy rain could dampen a storm's ability to form tornadoes. HP also can be a sign that the storm is playing itself out.

"If it's still producing hoses down by Cyril and these TV people are seeing them, it can't be wrapped up too bad," Roger said.

As they crested one rise and then another, the base of a supercell occasionally came into view. "Holy crap, Rog. It's got a lot of rotation under there. Get moving. Haul ass."

"I saw it. I'm trying, son, but I ain't going to drive like Richard Petty. We don't see anything at all if we wreck."

The Meatwagon crested a ridge just as the storm produced its sixth tornado. For the first time they saw Storm A, as it later would be labeled, in all its glory—its big, ragged wall cloud and its rotating mesocyclone so huge, so dark, and so menacing that it reminded Roger of the alien ships from the movie *Independence Day*.

"I think there might already be a tornado under there," Rich said.

Their arguing ceased, and they stared at Storm A's underbelly. The wall cloud sloughed off vertical columns of wind. Threads of multiple vortices began to twine and move together, forming a small cone-shaped tornado. It appeared to alternately shrink and expand a couple of times before growing into one fat cylindrical twister.

Because of the rolling hills and scrub oaks, they could not see the tip of the tornado on the ground. But they saw a small fireball explode near the ground. They thought the vortex had cut some electrical wires. It had in fact clipped a house.

Roger parked beside the road, and the two clamored out of the car to set up their video equipment. The supercell kicked out a long, lean satellite twister that orbited the main tornado. Now there were two funnels on the ground, the smaller tornado circling the larger one.

"We had never seen anything like that before. That's when we knew this was a special day—meteorologically and otherwise," Roger said.

The two forecasters marveled at the supercell and its offspring, captivated by the perfect symmetry of it all. The storm's mesocyclone was more than six miles wide, rolling along like a big wind-powered generator, inhaling the warm updraft and sloughing off a cool downdraft to its rear. At dead center, rotating ever so furiously, was the tornado—a grand, powerful, beautiful thing to behold.

The supercell and its twin tornadoes moved toward the northeast at a slow, steady pace. Oklahoma City was fifty miles dead ahead on the supercell's angled pathway, but neither man imagined that the storm would remain so productive. They also realized that this supercell wasn't going away anytime soon.

"Damn," Roger shouted to Rich over the roaring winds. "This storm's crankin' big time."

3

A Tornado Forecast

ON MARCH 20, 1948, TORNADO ALLEY introduced itself to U.S. Air Force Captain Robert C. Miller. A twister forced Miller to hug the floor where he crouched for safety amid the radar scopes and his hand-tinted forecast maps that were, it was clear to him now, woefully inaccurate. There was a terrible racket outside as a jumbo tornado mowed through a line of mothballed bombers, exploding windows, ripping off siding, and appearing to kill a fledgling meteorologist's career.

A native of southern California, Miller had little knowledge of tornadoes. In World War II, he was assigned as a weather forecaster to Dutch New Guinea, a South Pacific island. He knew a lot about sunshine, interrupted by periods of rain, followed by more sunshine. After the war and a brief interlude in Georgia, he found himself assigned as a forecaster to the Air Weather Service at the relatively new Tinker Air Force Base near Oklahoma City.

The war was nearly over by the time Tinker's construction finished. Oklahoma City leaders staged a pitched lobbying effort to woo the Air Force to a level stretch of pastureland southeast of the city. The Air Force was searching for a suitable site in the central United States for a maintenance and supply depot. Apparently no one mentioned anything about the weather. Oklahoma City won its lobbying effort, and the new base was named in honor of Major General Clarence Tinker, an Oklahoman and World War II pilot killed during a strike on Wake Island early in the war. During the war, the Tinker crews repaired B-24 and B-17 bombers. After the war, surplus bombers lined Tinker's perimeter, all in a line awaiting their final disposition, which for many would come in the form of a tornado.

On that March evening, Miller worked the late forecast shift. He had been in Oklahoma for only three weeks. The captain chatted up the staff sergeant on duty, and the two learned they had much in common: a southern California background and zero experience in forecasting Great Plains weather.

"We analyzed the latest surface weather maps and upper charts and arrived at the sage conclusion that except for moderately gusty surface winds, we were in for a dry and dull night," Miller said. "This forecast gravely underestimated the gravity of the situation."

Neither the captain nor the staff sergeant had noticed that the upper-air analysis sent to them via facsimile from the U.S. Weather Bureau in Washington was wrong. The Weather Bureau failed to accurately note the moisture content in the air. Because of the error, the Air Force and civilian forecasters underestimated the volatility of the atmosphere and failed to note the possibility of thunderstorms.

By 9:00 p.m., weather stations to Tinker's west and southwest began reporting lightning. A half-hour later, a full-fledged thunderstorm was in progress twenty miles from the base. Even milky images on Tinker's weather radar—a B-29's surplus radar scope—appeared vicious. Will Rogers Airport, Oklahoma City's main airfield west of Tinker, dispatched a warning of wind gusts of ninety-two miles per hour and ended the message with "Tornado South on Ground Moving NE!"

The sergeant began typing up a warning to alert Tinker personnel, but it was too late to secure the aircraft. At 10:00 p.m., lightning flashes allowed Miller to see the twister as it crossed over the air base. "We watched it, not really believing, as it passed just east of the large hangars and the operations building where we crouched in near panic."

The glass windows in the control tower exploded, sending shards into the bodies of tower personnel. A large window in the operations building burst outward. Debris whirled in the air. The funnel left as quickly as it came. The real damage was to the unsecured bombers left in the open. The loss to Tinker was $10 million.

The $10 million loss so upset the military that it sent five generals from Washington, D.C., to Tinker the next day. Miller and his immediate superior, Major E. J. Fawbush, were called in for a grilling. Miller thought his career was coming to an end. "It really didn't seem fair that a bright young forecaster, native to an area where a mild thunderstorm was considered a holiday event that caused people to run outside and gesticulate skyward mouthing such phrases as 'golly' and 'wow,' should be thrust into an area subject to such miserable phenomena."

Self-deprecation aside, Miller was a skilled analyst. In the book *Scanning the Skies*, Marlene Bradford noted Miller could craft a prognostic chart with the best of them. He created his own method of sketching out data at different levels of the atmosphere on a clear acetate map of the United States. By layering one atop the other, he created a three-dimensional view of the sky.

Colleagues who worked with him recalled that Miller "was a highly skilled and practiced analyst. His final charts were both incredibly detailed and also were often beautiful works of art."

But at Tinker, before the inquiring board of officers, Miller tried to melt into the background. Fawbush did most of the talking, describing the difficulties in forecasting tornadoes and the reluctance of the Weather Service to warn the public. He skipped the part about the erroneous moisture data from the U.S. Weather Bureau, wanting to avoid a battle with the civilian agency.

By the early afternoon, the military deemed the March 20 tornado "an act of God" that "was not forecastable given the present state of the art." It also recommended that meteorologists consider methods for public alerts and asked air base commanders to develop safety plans to protect people and property from severe storms.

After the tribunal departed, General Fred Borum, Tinker's commander, summoned Fawbush and Miller to his office. Twice in less than a year, Borum had witnessed the destructive force of a tornado. He had been stunned by the severely injured Woodward victims evacuated by Tinker pilots eleven months earlier and now his own base was in tatters. Borum ordered Fawbush and Miller to investigate the possibility of forecasting tornado-producing thunderstorms.

For the next three days, Miller and Fawbush studied U.S. Weather Bureau charts and reports from tornado outbreaks. They were looking for the telltale clues that would allow them to predict an outbreak with some degree of certainty. Much work on developing forecast patterns for storms had been done by the Weather Bureau, but it had never taken the next step of actually forecasting a tornado. There was little knowledge of a thunderstorm's structure, and nothing was known about the mechanics of tornado formation. Technology was of little help. Forecasters had little more than thermometers, air pressure gauges, and anemometers, and all three tools had been around since the seventeenth century. The newest forecasting gadget was the radar scope, usually taken from a scuttled airplane. The radar allowed forecasters to see the outlines of storms, but not what was happening inside the clouds.

Miller and Fawbush looked for patterns, such as moisture levels in the air, wind speed and direction, rising and falling temperatures, and rising and falling air pressure, that were present during past tornadoes. They listed several weather parameters, such as dew points and wind shear, considered sufficient to result in significant tornado outbreaks when all were present at the same time in the same location. The difficulty was to take all those data, determine the parameters or probability of the parameters that would exist, and project forward with some degree of confidence. Predicting a thunderstorm was one thing; predicting a tornado was quite another.

On the morning of March 25, only five days after the initial tornado, Fawbush and Miller noted the same upper-air and surface patterns as March 20. They redid the analysis even though the odds of tornadoes' taking the same path five days apart were astronomical—Miller estimated the odds at 20 million to 1. Still, he and Fawbush prepared a chart that "resulted in the somewhat unsettling conclusion that central

Oklahoma would be in the primary tornado threat area by late afternoon and early evening." They notified General Borum.

Miller noted that Borum was knowledgeable about weather trends and "loved to watch the [radar] scope during thunderstorm outbreaks." Borum absorbed the news. "Are you planning to issue a tornado forecast for Tinker?" Silence greeted the question. Fawbush turned to Miller.

"Well, it certainly looks like the twentieth, right, Bob?"

Neither man wanted to edge onto this dangerous limb.

"Yes, E. J., it certainly looks like it did on the twentieth."

Much to the relief of the two officers, the general suggested they issue a severe thunderstorm forecast, an alert that would put in action his new weather safety plan. At midday, it was clear that at least a thunderstorm was headed their way. Thunder clouds were shooting up in North Texas and western Oklahoma. By 2:00 p.m., the storms were building, increasing, and organizing into squall lines. Fawbush and Miller estimated the line would reach Tinker about 6:00 p.m.

Borum again asked the two if they were going to issue a tornado forecast, but Fawbush and Miller could only agree that the conditions were strikingly similar to the previous week.

"You two sound like a broken record. If you really believe this situation is very similar to the one last week, it seems logical to issue a tornado forecast."

The two again tried to backpedal, pointing out the "infinitesimal possibility" of a second tornado striking the same area in twenty years, let alone five days. "Besides, no one has ever issued an operational tornado forecast," they said.

"You're about to set a precedent," the general replied.

Fawbush composed the message. Miller typed it up and at 2:50 p.m. on March 25, 1948, passed it along to base operations.

Miller pondered this career suicide: "I wondered how I would manage as a civilian, perhaps as an elevator operator. It seemed improbable that anyone would employ, as a weather forecaster, an idiot who issued a tornado forecast for a precise location."

At 5:00 p.m. there still was no sign of a heavy thunderstorm near the base. In fact, the winds were light. Miller went home and left Fawbush to deal with the fallout over their tornado forecast. An hour later, Miller was at home listening to the radio when the broadcast was interrupted by a news flash about a tornado at Tinker. He initially thought the announcer was updating the news about the twister five days earlier, and it took him a couple of minutes to realize something new had happened.

Miller rushed back to the base to find debris and utility poles littering it. At the weather station, he found a jubilant Fawbush. The major described how two thunderstorms appeared to join together just southwest of the base. The clouds turned greenish-black and began rotating at the merger point. He saw the wing of a mothballed B-29 float upward and then splinter apart as it touched the edge of a funnel. The second tornado caused $6 million in damage to the base, but there were no injuries. Most airplanes had been secured or placed in hangars.

Now the two men had to prove their method was no fluke.

It would be another year before Fawbush and Miller issued their second tornado forecast. During that time, they refined and tested the basic forecasting rules that would enable them to predict with some degree of accuracy the thunderstorm formations most likely to produce tornadoes. On March 25, 1949, a year to the day after their first forecast, Fawbush and Miller successfully issued a second tornado forecast for southeastern Oklahoma. Two tornadoes occurred near McAlester in the southeastern corner of the state.

They also worked with William Maughan, the chief meteorologist at the U.S. Weather Bureau office in Oklahoma City. Maughan provided the men with forty years of weather data, especially information on the Woodward tornado of 1947. According to *Scanning the Skies,* Maughan sought to interest his civilian superiors in the two Air Force officers' work but was rebuffed. Maughan did receive permission to forward the military tornado forecasts to the Oklahoma Highway Patrol and the American Red Cross, but he was prohibited from forwarding the forecast to the public.

A month later, Fawbush and Miller issued a third tornado forecast.

Maughan forwarded the Air Weather Service message to state troopers and the Red Cross. That day, thirteen tornadoes struck Oklahoma, killing six people. Encouraged by their accuracy, the two men expanded their forecasts to military bases around the region but were still barred from forwarding their predictions to the other Weather Bureau offices or the public.

The Tinker duo reopened an old debate with the Weather Bureau not only over the reliability of tornado forecasts but over who should be in charge of weather warnings: the military or civilians.

The Weather Bureau's roots dated to 1870, when the U.S. Army Signal Service had created a weather forecast office. One of its biggest backers was a shy, visually myopic astronomer named Cleveland Abbe. "The atmosphere is much too near for dreams. It forces us to action. It is close to us. We are in it and of it. It rouses us both to study and to do. We must know its moods and also its motive forces," Abbe wrote.

In 1869, Abbe, the astronomy director of the Cincinnati Observatory, convinced the Cincinnati Chamber of Commerce and Western Union to sponsor the daily collection of weather data from throughout the Ohio Valley. Using these data, Abbe made what he called probabilities, or forecasts, and provided a daily weather chart. "I have started that which the country will not willingly let die," he boasted to his father.

He gained a nickname, "Old Probabilities" or "Old Prob," and the gentle New Yorker became one of the best-known scientists of his era.

At the suggestion of Abbe and other weather enthusiasts, Congress established in 1870 a weather warning office within the U.S. Army Signal Service, later called the Signal Corps. Abbe, the only person with experience in forecasting weather, was appointed as the civilian assistant to the military Signal Service chief.

Abbe urged the War Department to study and research the weather so forecasters would have a scientific basis for their predictions. "From the start Abbe's plans met with obstruction at almost every turn, not always due to unfriendliness—indeed more often to mere inertia of the system. In overcoming this opposition, which at times was so unyielding as to completely discourage all others who were interested, he was successful, because two of his most characteristic traits were an inexhaustible enthusiasm for the work, which amounted almost to an obsession, and an equally inexhaustible patience in meeting unfriendly or unintelligent criticism," wrote his friend and Signal Corps colleague T. C. Mendenhall.

The military's orderly tradition and the civilian scientists' need for creativity failed to mesh. Mendenhall noted the chief Signal Service officer would send a regulation order slip to the scientists assigning them the duty, which Mendenhall jokingly parodied as an order to determine the cause of gravitational pull, of providing a preliminary report in two weeks and a final report in one month. The scientists were supposed to salute and follow the orders.

"That controversies between the two were on the whole rather infrequent and rarely acute was due, more than to anything else, to Abbe's unfailing good nature and general willingness to be the subject of the obloquy of both sides," Mendenhall wrote.

The job of the Signal Corps was to maintain military communications— telegraphs at the time. But the telegraph also enabled, for the first time, the communication of weather observations. At the time, the placement of the weather office within the Signal Corps made technical sense.

In 1877, a meticulous Army enlistee named John Park Finley, a bearish six-foot-three Michigan native, had joined the Signal Corps. Finley's interest was severe storms, especially tornadoes. Finley was assigned to the Fact Room, which created and published the *Monthly Weather Review* and the *Weekly Weather Chronicle* that Abbe edited. He also served as a damage surveyor for the Signal Corps. By horse and buggy, he traveled to the Central Plains to examine the results of a tornado outbreak in 1879. He traveled 500 miles through Kansas, Missouri, Nebraska, and Iowa, mostly during still-stormy conditions. He interviewed eyewitnesses, studied the direction of the winds based on felled trees, and collected weather data.

"The roaring of the storm was so great that the family, while in the cellar, did not hear or know of the destruction that was going on above them," he wrote of one family. "The roaring was so great, that they could not hear the screaming of their daughter for assistance in an adjoining room, even though the door between the two rooms was open," he wrote of another family.

He counted forty-two dead from the tornado outbreak. "All of the parties were covered with mud from head to foot; eyes, mouths, and ears filled, and clothing torn to shreds. The mother and two children were left in the rubbish; the former having her head crushed and her long hair, which reached below her waist, was partly cut and pulled from her head, twisted into a rope and found several feet from her body. That portion of her hair left upon her head was twisted into little wisps and mixed with mud. The bodies of the children, after having been washed for days, were still covered with specks of fine dirt and leaves which seemed to be driven into the flesh."

Finley noted the horror's lingering impact on the survivors: "The effect upon the people was pitiful in the extreme. Night after night hundreds of people never went to bed, but remained dressed and with their lanterns trimmed, watching for a fresh onslaught which they expected momentarily. Every dark cloud or sudden increase in the velocity of the wind seemed to fill them with evil forebodings which could not be allayed until every vestige of supposed danger had vanished."

What Finley lacked in terms of scientific knowledge about the tornado he recouped in terms of sheer doggedness for gathering facts. He obtained the times of various tornado occurrences at different locations and tried to calculate the speed of the twisters. No detail, from the description of the clouds to the changes in air pressure, was too small.

The Army private, with little proof except a hunch, made his own deductions: "As an area of low barometer advances to the Lower Missouri Valley, warm and cold currents set in towards it from the north and south, respectively. Warm and moist regions emanate from the Gulf and the cold and comparatively dry air from regions of the British Possessions. The marked contrasts of temperature and moisture, invariably foretell an atmospheric disturbance of unusual violence, for which this region is peculiarly fitted." Finley's description fit the stormy weather ingredients that would plague the Central Plains.

He also urged the U.S. Signal Corps to establish a weather office in Kansas City that could telegraph warnings throughout the Missouri Valley.

Finley became convinced that tornadoes could be forecast. A change in Signal Corps leadership brought General William B. Hazen to command, providing a measure of relief to both Abbe and Finley. Hazen, a supporter of weather research, established the Study Room, a research unit to which Finley was assigned. From the Study Room, Finley began to organize a network of observers, or reporters as he called them, to telegraph basic weather data. Eventually he would organize more than 2,000 reporters from the East Coast to the Rocky Mountains.

He also completed a study he had worked on for years, tracking reports of tornadoes dating back to 1794. Finley's 1882 publication, *The Character of 600 Tornadoes,* was the most comprehensive survey at the time. The same year, he married and enrolled as a graduate student at Johns Hopkins University to study tornadoes. It was all too much for Finley. Overwhelmed by his numerous duties, he was hospitalized for exhaustion. "He is doing too much brain work," the doctor wrote to Hazen. "He very decidedly needs rest. Should he continue to do as much as at present the consequences will be of very grave character."

The hospitalizations slowed him only slightly. Promoted to lieutenant, he resumed his tornado investigations and began experimenting with tornado predictions. He claimed a 96.6 percent accuracy rate, a record achieved by also counting as a success his predictions that a tornado would not occur.

On October 4, 1885, *The New York Times* ran a small item under the headline "Predictions of Tornadoes": "The phenomena of tornadoes, to the scientific study of which Lieut. John P. Finley, of the Signal Corps, has devoted about eight years, are now so well understood as to warrant the belief that trustworthy warnings may soon be sent out to inhabitants of localities which may be threatened with disastrous visitations," the paper said. It noted the forecasts were experimental and that daily bulletins would note when conditions were favorable for the development of a tornado. "In such cases 'severe local storms' are noted as probable."

The forecasts were short-lived. In 1887, nervous superiors sent him new instructions: the word *tornado* was banned from his forecasts. He was ordered to refer to "severe local storms" rather than use the word *tornado*. Other weather officers received the same order. There were economic reasons. Businessmen in Iowa and other territories complained that Finley was giving potential investors the idea that their region was twister prone.

The U.S. Signal Corps was in the middle of the political dispute. Should the weather service be in the hands of the military or civilians? Budget cuts, scandal, and politics combined into a two-year investigation by Congress. Finley was one of four officers admonished for mistreating subordinates.

A congressional commission recommended the weather operations be transferred from the Army to the War Department. The Signal Corps training and research centers were closed. Finley also lost support from his new commander. General Adolphous Greely, the new chief signal officer, chided the idea of tornado forecasts: "It is believed that the harm done by such predictions would eventually be greater than that which results from the tornado itself." The government would maintain that position for sixty years.

In 1890, Congress created the U.S. Weather Bureau within the Department of Agriculture. Abbe, who stayed with the government until his death in 1916, became its chief meteorologist and continued to press his case for more emphasis on meteorological studies as a science. (He also successfully pushed for the creation of standard time zones based on Greenwich Mean Time.)

Finley was detailed briefly to the Weather Bureau but quickly rejoined his regiment and gave up meteorology, at least for a while. He retired from service in 1914 and began a second career as a private meteorologist and risk assessor to insurance companies. Even late in life, Finley could spot a weather trend. At age seventy-eight, he started the National Weather and Aviation School in Michigan, astutely connecting the future of weather forecasting with a rising new industry: commercial aviation. He was ahead of his time once again. Aviation eventually would spur the Weather Bureau into action. He died in 1943, not living long enough to ever see or hear another official tornado forecast.

Under the Department of Agriculture, the Weather Bureau began to issue flood warnings, created the nation's first daily weather map, established a hurricane warning network, set up an information exchange with Europe, and issued fire weather forecasts. In 1909, a justifiably proud Abbe wrote in *The New York Times* of the advancements made in the past fifty years: "Formerly the instruction in meteorology in both schools and colleges amounted to nothing. But now the daily weather map has become our textbook and elaborate studies of it and what it teaches are carried on in every high school and many colleges from Tacoma, Wash., eastward to Cambridge, Mass., and southward to New Orleans. In fact, meteorology has become established as a branch of sciences and it is properly called 'the physics of the atmosphere.' It is no longer the mere description of appearances but a study of the forces at work in the atmosphere to bring about these appearances. All nations now maintain weather bureaus and atmospheric researches. Nature holds her secrets in an iron grip, but we are wresting them from her."

It would take World War I to focus attention on weather studies, but this time the cutting-edge discoveries came from Europe. Meteorology was about to emerge from a theoretical backwater into a full-fledged science. At the University of Bergen in Norway, Vilhelm Bjerknes and his son, Jacob, provided a critical breakthrough for practical weather forecasting. The Norwegians discovered atmospheric oceans of air they called masses. The boundary zones between the warm and cool air masses were called fronts, after the trench warfare battle lines of World War I. The new theory was critical to understanding the atmosphere and forecasting weather events.

What seems so obvious now was revolutionary at the time—so revolutionary that U.S. scientists and the Weather Bureau dismissed the "Bergen school" air mass theory for nearly two decades. In the United States, there would be no discussion of warm and cool fronts, no thought given to air masses, and no credence given to what might happen when those two opposing masses collided.

The consequences were enormous.

On March 18, 1925, the Weather Bureau predicted "rains and shifting winds" for Missouri, Illinois, and Indiana. It would be the agency's greatest blunder.

Survivors recall the sky turning day into night as it darkened with the brewing thunderstorm. About 1:00 p.m. near Ellington, Missouri, it produced a tornado so massive, so destructive, and so deadly that it still holds the record for the largest killer twister ever.

From Missouri through southern Illinois and into Indiana, the Tri-State twister traveled 219 miles in three and a half hours, moving at more than twice the speed of a usual tornado. It overran one town after another. The only warning was whatever citizens saw coming at them.

It leveled farmhouses, neighborhoods, schools, and office buildings. By 4:30 p.m., when it dissipated, the dead totaled 695. It remains the deadliest tornado in history.

Throughout the 1920s and 1930s, the death toll from tornadoes mounted. There were no warnings, no time for people to seek shelter. On April 5, 1936, a twister in Mississippi killed 216 people in Tupelo. The next day, the same violent storm system spawned another tornado that killed 203 people in Gainesville, Georgia.

In 1938, as fatalities rose, the Weather Bureau lifted its ban on the use of the word *tornado* but mainly in its alerts to emergency personnel, not to the public. No forecaster ventured a public tornado prediction.

Bradford noted: "Although weather observation and data collecting had been around for centuries, during this period meteorology was a relatively new science, and most Weather Bureau personnel had learned their craft through on-the-job experience rather than in a college classroom. Only in the mid-1930s did the bureau adopt the frontal theory of storms and hire university men trained in the use of this new forecasting method. . . . In spite of advancements in meteorology and technology, a system of tornado forecasting and warnings was as nonexistent in 1940 as it had been in 1870."

In 1940, Congress again reorganized the U.S. Weather Bureau and placed it under the Department of Commerce, a nod toward the burgeoning importance of climate and weather to the nation's economy.

During World War II, the Weather Bureau did muster up a spotter network to help protect munitions plants and strategic factories. Radio-equipped volunteers positioned themselves around thunderstorms to provide warnings to the plants and factories should a tornado appear.

The thunderstorm increasingly posed a threat to the nascent civil aviation industry, as Finley had predicted it would. A series of storm-related airplane crashes prompted Congress to approve and finance the first official study of thunderstorms.

The 1945 Thunderstorm Project was headed by Horace Byers, a leading meteorological researcher based at the University of Chicago. With the war over, Byers and his colleagues finally accessed surplus airplanes, crews, and equipment needed to conduct a massive research project. They also had a new military tool used during the war: the radar.

"We will study and probe the thunderstorm in much the same way that a zoologist studies a new organism," Byers pledged to Congress.

And probe they did. Byers and his team used Northrop P-61s, stiff-wing fighters dubbed the Black Widow, to fly simultaneously into thunderstorms at different elevations. The twin-prop P-61 had been designed as a nighttime fighter, its long nose cone filled with radar equipment. The first flights took place in Florida in 1946 because of the state's propensity for thunderstorms. The next year, the flights took place over Ohio, near an Air Force base that provided operational support.

The Black Widow pilots had to fly into a storm stacked five high, one at every 5,000 feet of elevation. They also had to fly into the storm straight and level and try not to touch the controls as winds jostled the tiny fighters. The idea was to measure the turbulence and winds within the storm.

Perhaps the most immediately beneficial discovery from the Thunderstorm Project was that radar, which was top secret during the war, could be used aboard airplanes to avoid thunderstorms altogether. Pilots could simply fly around the storms rather than try to bull their way through the tempest.

The Thunderstorm Project provided the initial outlines of the structure of a thunderstorm. Scientists were able to document a storm's life stages: birth, maturity, and dissipation. They also noted the existence of a cool downdraft.

Fujita's model of *Raiu-no-hana,* or Thunder-Nose, displayed his theory of air being sucked into the front of the storm and discharged out the rear. Like Byers, he viewed the thunderstorm as an individual organism. He called it microanalysis. Later, in the United States, it was called mesoanalysis, and it provided a fundamental new model for studying the behavior of an individual thunderstorm.

Fujita made his downdraft discovery using only pencil and paper and the most basic of weather equipment. Byers was impressed by Fujita's downdraft theory, and the two began an exchange of letters and research papers. "In view of the fact that you were not familiar with the work of the U.S. Thunderstorm Project on this subject your conclusions are highly valuable and really represent an independent discovery of some of the factors derived from our work," Byers wrote him.

The 1948 tornado forecast by Fawbush and Miller only added to the Weather Bureau's embarrassment. In 1950, the Weather Bureau denied that it had a policy of banning the word *tornado.* Francis Reichelderfer, Weather Bureau director, issued a circular to local offices advising them to avoid phrases such as "The Weather Bureau does not make tornado forecasts" or "We are not permitted to make tornado forecasts" when talking to the media. Reichelderfer said such statements "incorrectly implied that the bureau was unwilling or unable to make tornado forecasts." The note also warned, however, that because tornado forecasting was one of the bureau's "most difficult tasks," a "good probability of verification" should exist when such forecasts were made. In other words, there would be hell to pay if a tornado forecast was made and one did not occur. No Weather Bureau forecaster was going to go out on that limb. The probability of failure weighed more heavily than success.

Even the American Meteorological Society (AMS), which reflected the cautious nature of the Weather Bureau when it came to tornadoes and forecasting severe storms, was highly skeptical of the Air Force's approach. It agreed to hear a presentation by Fawbush and Miller during its 1950 annual meeting, to be held in St. Louis. But it would hear them only behind closed doors. Neither the press nor public was to be allowed to hear the Tinker meteorologists' presentation.

Nature intervened. An off-season tornado hit East St. Louis the day before the AMS meeting. Tinker officials sent a news release to local reporters telling of the Air Force's tornado forecasting successes in Oklahoma. It made front-page news. The AMS was forced to open its doors, and the press made headline news of the presentation by Fawbush and Miller.

Intrigued by the work of Fawbush and Miller, the Air Force in 1951 established the Severe Weather Warning Center (SWWC) within the Air Weather Service at Tinker. The two meteorologists were placed in charge, responsible for all severe weather warnings to all Air Force bases throughout the United States.

Forecasting severe weather and recognizing the subtle patterns that made conditions ripe for a tornado required much time and analysis. Creating the SWWC gave Fawbush and Miller time to focus on severe weather data and freed meteorologists at other military installations to concentrate on daily forecasts. The SWWC routinely provided its forecasts to the Weather Bureau offices and emergency personnel in affected states, but the courtesy only served to embarrass the bureau further.

The Weather Bureau's Washington managers complained that Fawbush and Miller were intruding on the agency's territory by forwarding their forecasts to local offices and law enforcement. The military stopped releasing its forecasts to anyone but its own personnel.

The Weather Bureau won the argument that it should be the sole source of civilian weather forecasting and public warnings, but it refused to issue such warnings. Its public battle with the Air Force served only to highlight the military's successful tornado forecasting program and increased pressure on the Weather Bureau either to pass along the military's warnings or issue its own.

In Oklahoma, the media were in full howl against the Weather Bureau and Reichelderfer. The Oklahoma media were well aware that the

Tinker meteorologists were routinely issuing tornado forecasts. Newspapers and radio broadcasters accused the Weather Bureau of jealousy, obstinacy, and incompetence. The idea that the SWWC could warn military bases but the Weather Bureau would not alert the public of impending tornadoes smacked of plain stupidity.

Reichelderfer, in turn, stewed not so quietly about the dubious accuracy rates claimed by Tinker. The Air Force meteorologists had the luxury of repeatedly fine-tuning their forecasts during a weather outbreak because they could easily contact the affected military base. The Weather Bureau, in contrast, had to stand by its forecasts once the mass media were alerted. It could not repeatedly make forecast refinements because its audience was too large.

A former Navy weatherman, Reichelderfer had studied meteorology at Harvard and at the famed University of Bergen in Norway, and he wanted hard science behind the forecasts. The Tinker team used the most basic form of forecasting, pattern recognition. But the dynamics of a thunderstorm and the whys, wheres, and hows of tornado formation were all but unknown.

Politically, Reichelderfer could no longer maintain his position.

The Weather Bureau's forecasters based in Washington, D.C., made their first experimental warning in March 1952, issuing a tornado forecast so late and so quietly that few knew of it, which was just as well because it was inaccurate. The next evening, they tried again, issuing a tornado warning for seven states between Texas and Indiana. Thirty-six twisters struck in almost every state. Whatever jubilation they may have felt was tempered by the number of fatalities. The dead totaled 208.

Reichelderfer chose five men to become part of the new Severe Weather Unit. The unit was renamed Severe Local Storm Unit (SELS) when it moved to Kansas City, just as John Finley had suggested a half-century earlier. Tinker's weather operations, also renamed the Military Weather Warning Center (MWWC), previously had moved to Kansas City because of its central location. The competition would continue for years.

A Tornado Forecast

The stogie-chewing Miller was extremely competitive and savored every chance at besting SELS with a forecast.

Military meteorologists who would later make their careers on the civilian side learned their trade from Miller. By now a colonel, Miller personally selected the men who would serve at his group and chose those with a passion for severe storms.

"The high point of an MWWC forecaster's day would be to come in and find a huge 'Great!' marked on your forecast chart boldly in large purple print. The opposite experience of finding his equally bold critique of a bad forecast was always disheartening and embarrassing in the short term. Seeing the comment of 'Dummy! How could you have missed this #$#%' in bold purple for all to see was tough," recalled a coworker.

The public would never know that Miller's military operation was kicking the civilians' tails. But the civilian forecasters certainly knew, and the Weather Bureau began to assign younger and college-educated men to the team—people without preconceived notions that a tornado forecast was futile.

In 1953, a series of deadly tornadoes struck Texas, Michigan, and Massachusetts. On May 11, 114 people died in Waco; on June 8, 115 were killed in Flint; on June 9, 94 people were killed in Worcester. Only the weather office in Massachusetts failed to offer any weather warning, convinced that a tornado could not possibly develop in the state.

The year was one of the deadliest since the 1925 Tri-State twister. The 1953 death toll rose to more than 500, due in part to the lackluster warnings but also because of the strength of the furious twisters. The 1953 triple tragedies prompted some members of Congress to question whether the nuclear weapons testing in Nevada was affecting the weather. The weathermen discounted a connection.

"If there were hydrogen bombs exploded every day for several months, or for a year or two, it would have an influence on our climate, but I think our worries would be other than climate if there were that many hydrogen bombs exploded," Reichelderfer testified before Congress.

STORM WARNING

The year 1953 would prove to be a benchmark for weather and weather research. Not since 1953 has one tornado killed 100 people or more. U.S. Weather Bureau tornado warnings, the rise of television, and even the fear of nuclear war would coincide to bring a modicum of relief from the deadly twisters.

And in 1953, Tetsuya Fujita arrived in America.

4

SEARCHING FOR CLUES

A T MIDNIGHT, ROGER EDWARDS SLIPPED THROUGH a side door at the Storm Prediction Center toting a gym bag containing a two-liter bottle of Diet Dr Pepper, a covered bowl of leftover dinner, and a clear plastic tackle box filled with colored pencils. Roger worked the graveyard shift on May 3. The midnight-to-8:00-a.m. duty was Roger's favorite; it was quiet, the building was empty except for three meteorologists on overnight, and it gave him time to work on research projects that usually involved some aspect of tornado mechanics.

Three forecasters were the minimum on staff during any shift: the lead forecaster who oversaw all forecasts, the mesoscale forecaster who looked for the prospects of severe storms anyplace in the United States within the next six hours, and the outlook forecaster who plotted the chases of severe or nonsevere storms within the next three days. Roger was working as the outlook forecaster, writing the Day 1 Convective Outlook, which served as the morning guide for the National Weather Service.

From a keyboard, Roger toggled between three computer monitors that could display the nation's most advanced weather detection system. Here, sitting in a sleek ergonomic chair and under a harsh glare, Roger could access information from 116 Doppler radars and three stationary satellites above the United States. Wind direction and speed, moisture and air pressure from hundreds of locations, and numerical models spewed from one of the world's most advanced meteorological super-computers.

But those colored pencils were his favorite tool.

His first act, by habit and necessity, was to sketch by hand the pressure lines, the highs and lows, taking careful note of the changes in the previous twelve hours. He might do one or several U.S. maps, each paper copy presenting data from different altitudes of the atmosphere. He studied the wind speed and directions and the temperature and pressure levels that marked the position of various weather fronts. Roger kept his own pencil cache because sometimes lime green or fuchsia struck his fancy.

"You can't predict what the weather is going to do until you know what it is doing," he explained.

He used a "K" for cold spot, a nod to *kald*, the Norwegian word for cold. It was the work of the Norwegians that created the modern weather map, based on the battle schemes of World War I. The standardized weather map included boundaries between cool and warm air fronts. Cold fronts were marked with a line of triangles reminiscent of spiked World War I German helmets and the warm front with the rounded half-circles similar to British helmets. Much of the language of meteorology related to warfare.

Roger drew the isobars, the curvy lines that delineated the high- and low-pressure systems, and used blue pencils for the cool fronts and red for warm fronts. He liked this old-fashioned, hand-drawn method—"the right way," he deemed it, the way Miller, Fawbush, and the Tinker meteorologists used to do it. A computer could do it instantly. But drawing the atmospheric conditions by hand gave him an intimate feel for the weather that a computer rendering never could. "Forecasting is as much an art as it is a science," he said. "This is the art part." He finished by writing his name at the bottom of the page and made a swiggly little vortex beside it.

Sometimes a forecaster must go with instincts and experience. Even the supercomputer used by the Storm Prediction Center (SPC) is fallible, although its conclusions are rarely as imperfect as they were on May 3, 1999.

"It may sound trite, but by far, the most important software in the tornado forecast process is within the human brain. The forecaster must use it to sort all that information, toss out what is not needed, properly interpret what is needed, and put it into a coherent form—all on a time deadline," Roger wrote on the SPC Web site explaining a forecaster's job.

On May 3, the pieces fell together ever so slowly as Roger sketched the clues across a map. He looked for moisture, instability, lift, and shear—the building blocks for a severe storm. A trough, an elongated low-pressure system, positioned itself over the high plains of northeastern Colorado, spinning the nearby air counterclockwise. Pulled by the low-pressure system, a tongue of warm moist Gulf air jutted across Texas and into central Oklahoma. It was obvious to him that there would be a thunderstorm, but nothing appeared out of the ordinary.

"It looked like a fairly middle-of-the-road, severe-weather day in May for Oklahoma, which usually means hail, wind, and maybe a couple of tornadoes. If you have a severe weather day in Oklahoma, that's the average potential. That's what we went with most of the night," said Roger.

Forecasting is an effort to answer as many "maybes" as possible with as much confidence as possible. On May 3, the maybes lasted until the first storm initiated.

At 4:00 a.m., radar and wind profilers indicated a burst of middle- to upper-level winds across the California and Nevada borders, the fast, high winds of a jet streak. West Coast balloon-borne soundings from 7:00 p.m. the previous evening failed to detect the jet current as it came ashore from the Pacific Ocean. Now the winds aloft were barreling across the deserts at ninety knots, picking up speed as the ocean moisture evaporated. The big supercomputer crunching millions of pieces of weather data from its home base in Maryland also failed to account for the wind. In a science where the merest ten-thousandth of a fraction looms large, the failure to adjust to the powerful jet streak was a huge error.

The three SPC meteorologists pondered these new data. How would the jet streak affect wind shear over Texas, Oklahoma, and Kansas twelve hours later? Where were the storms going to form? Moisture wasn't a problem; plenty was coming in from the Gulf. Instability and shear weren't a problem; the low-pressure system over Colorado was swirling the atmosphere, and the jet streak would soon join the mix. They considered whether the winds would force thunderstorms into a squall line, which reduces tornado potential, or whether it would give life to individual thunderstorms and supercells, which would increase the tornado potential.

Usually in May, the dryline boundary between the Gulf and Rocky Mountain air currents was definitive; the winds on either side go in opposite directions and help fuel storm initiation. But on May 3, the dryline boundary appeared defused, with winds on either side moving in the same direction.

Roger and the SPC meteorologists began recalculating the data to include the strong upper-level wind profiles. All four computer models agreed there would be precipitation, but there was no agreement on a location—not even close. Each numerical model forecast rain in different parts of the region. Another unknown was lift. A warmed surface meant heat thermals jettisoning upward. These vertical thermals lifted the air into the atmosphere, creating a fuse for thunderstorms. From the satellite pictures, Roger could see a thin layer of cloud cover developing over West Texas and moving toward Oklahoma. The cloud cover could block sunshine to the surface and reduce surface heating, which could reduce the lift needed to push the warm air upward.

The SPC awaited data from morning balloon soundings. Twice each day at 0 and 1200 Greenwich Mean Time—Zulu time, or Z for zero, to meteorologists—weather stations worldwide sent aloft six-foot latex balloons with radio transmitters attached. In Norman, National Weather Service forecasters released their balloons at 7:00 a.m. and 7:00 p.m., central daylight time.

Attached to the balloon was a radiosonde, a minitransmitter that relays measurements for temperature, wind speed and direction, air pressure, and humidity at different levels of the atmosphere. The radiosonde provides meteorologists an idea of what is happening in the upper levels of the atmosphere. As it rose, the balloon expanded because of decreasing air pressure, eventually bursting and falling back to Earth.

Roger conferred with forecasters at the Norman office of the National Weather Service, located across the street from the SPC. "We had a lot of doubt as to where or how storms could form. There was not an obvious boundary. We didn't know exactly where those storms would form," said Roger.

At 8:00 a.m., Roger issued a "slight risk" for thunderstorms for Texas, Oklahoma, and Kansas based on the obvious levels of moisture and instability. "Slight risk" is the first of three SPC threat levels: slight, moderate, and high. Roger considered issuing a moderate risk, but neither he nor the other forecasters could pinpoint an exact location for severe weather. This ambiguity would dog meteorologists throughout the day. The morning forecast from the Norman office of the National Weather Service highlighted the probability of thunderstorms and the possibility of a tornado later in the day.

An SPC weather watch, a parallelogram that covers about 25,000 square miles, had consequences. It activated the storm spotter network operated by the National Weather Service, put local emergency managers on notice, signaled airports of possible aviation hazards, and provided the first hint of possible thunderstorms or tornadoes to the media and the public.

Once a storm or tornado was spotted or appeared imminent, the local National Weather Service office issued a severe thunderstorm or tornado warning. A warning was more serious than a watch. As defined by the government, a warning must be issued if a severe thunderstorm has at least ¾-inch hail, winds of 58 miles per hour, or a tornado. National Weather Service warnings immediately were transmitted over the NOAA Weather Radio and to the media. Only the National Weather Service could issue an official warning. A tornado warning meant people should seek shelter immediately.

As Roger left for home, a soft southern breeze blew across Max

Westheimer Field. The skies were partly sunny, with blue breaking through a few wispy thin cirrus clouds. It looked as if it would be a nice day, but he would miss it; he was going home to sleep.

BEFORE DAWN, AS ROGER pondered the meaning of the newly discovered speedy jet streak high over the California desert, Kara Wiese's eyes fluttered open and she inched her way out of bed. It was Monday morning, a workday for her at Financial Capital. She worked in the billing office, assigned to call small businesses that were behind on their loan payments.

The morning paper called for a partly cloudy day, highs in the 70s, but the forecast had changed overnight based on SPC and National Weather Service outlooks. Already television and radio stations were warning viewers of the probability of rain and the possibility of severe weather.

Kara arrived at work as early as her employer would allow, usually around 7:30 a.m., because it gave her time in the early evenings to do something—play catch, watch T-ball—with Jordan. But her early schedule also meant she had to roust a sleepy six-year-old from his bed and get him showered and fed before dropping him off at day care near the Bridge Creek School. The day-care center would drive Jordan back and forth to his half-day kindergarten class at Bridge Creek, so it was one less thing Kara had to worry about. Jordan would enter first grade in the fall, but he already knew his numbers and the alphabet. Still, it was a long workday for both of them.

Kara and Jordan had spent the weekend around her family. Her brother, Dustin, was getting married in June. Kara and Jordan; her mother, Mary; Dustin; and his fiancée, Misty, spent much of Saturday together shopping for clothes and cars. Kara was the first to spot a red sports car and insist that Dustin had to buy it as the new family car. He did, long ago having given up arguing with her. The two had more than their share of sibling spats; he found her bossy, and she found him immature. But in the past year or two, they had become much better friends. "I grew up," explained Dustin, twenty-one at the time. Besides, when Kara set her mind to something, it was best to agree early. At the end of their shopping spree, they found themselves in a Taco Bell for dinner. And it was here, in this Taco Bell, over beans and burritos, that they would all realize that at that exact moment, they were all very happy.

It had been a difficult few years for Kara, ever since her cousin and best friend, Rhonda, died in a car accident. Kara had been pregnant with Jordan. The cousins had been the mainstay of the extended family. It was Kara and Rhonda who planned the picnics, the anniversaries, the Christmas celebrations, the graduation parties. They were strong-willed and opinionated and usually joined at the hip.

Rhonda's death pushed Kara off-kilter, it seemed to her mother. Kara finished her junior college studies in accounting, but scrapped her plans to get a university degree. A couple of years later, she married when she probably shouldn't have. She had qualms but kept them to herself. Kara's new husband talked her into selling her small house in south Oklahoma City and moving into a mobile home at Bridge Creek. It didn't take long for her to ditch the husband; the marriage lasted six months. But she kept the trailer. She liked Bridge Creek. It was an easy commute to her job, and there was a school nearby for Jordan.

Finding Kara's place, or even Bridge Creek itself, had become a family joke. Mary had never heard of Bridge Creek until Kara moved there. Bridge Creek didn't rate even a pinpoint dot on the state maps. Tucked into the northern corner of Grady County, wedged between the small towns of Newcastle to the east and Tuttle to the west, Bridge Creek was burrowed deep into the countryside, dissected by a maze of unpaved section-line roads. Interstate 44 served as its eastern boundary.

"It took me several trips after she moved to Bridge Creek before I felt like I was not going to get lost," recalled Mary. The first time Dustin used his sister's convoluted instructions, he ended up lost fifteen miles away from her house.

For Mary, it was finally time to relax; her children were grown and happy. Dustin was getting married, and Kara seemed back on track.

Kara had worked hard to get her finances back in order and had announced she intended to return to college that fall. She wanted to get a degree in accounting. And she had met a new man, Lee, and everyone liked him. But Mary also knew Kara was thinking about Rhonda and death because they had talked about it over the weekend. Kara had been having dreams again, dreams about Rhonda. The car crash had left Rhonda in intensive care for six weeks before she died. And the anniversary of her death was approaching; it was Monday, May 3.

IN THE PARKING LOT of a Super 8 Motel, Craig Wolter quickly counted the heads of all twelve sleepy high school students. Wolter taught science at Windom High School in Minnesota, and his class on meteorology was his passion.

They climbed into two white school vans and headed out for the Storm Prediction Center. Wolter was as excited as the students. "I think these guys walk on water," he said of the scientists there.

On Sunday, the previous day, the students had spent twelve hours in the vans as they drove the 700 miles from Windom, in southern Minnesota, to Norman—twelve hours across the flattest terrain in the country.

This was the second time Wolter had brought his top meteorology students to Norman. An amateur storm chaser, Wolter had attended an American Meteorological Society summer conference in 1997 and met Joseph Schaefer, director of the Storm Prediction Center. Schaefer invited Wolter to bring his students for a tour.

In 1998, Wolter and twelve students made the day-long drive to Norman, only to discover the next day that all the tornado activity was in Minnesota. "Every researcher there pointed out to us that we were in the wrong place at the wrong time."

They were about to be in the right place at the right time.

At midmorning, wind radar in Tucumcari, New Mexico, detected the speedy little jet streak that had been over the California-Nevada border. It became increasingly clear that the jet streak would add to the chaotic changes in wind speed and direction, or wind shear.

At 11:30 a.m., the Storm Prediction Center elevated its outlook to a moderate risk for severe storms and mentioned the potential for violent tornadoes. A more definitive location for the storms continued to elude them. The computer models, the dryline, the ingredients—all remained in flux.

Wolter and his students planned to meet with forecasters from the SPC and researchers from the National Severe Storms Laboratory and then trek over to the University of Oklahoma to meet with tornado-chasing scientists who operated the Doppler On Wheels (DOW) program.

At SPC, the Minnesotans were ushered into a darkened room already crowded with forecasters and researchers poring over the latest weather data. Everyone was trying to decipher the weather prospects. Rich Thompson was working the day shift as the mesoscale forecaster. Part of his job during severe weather was to create the mesoscale discussion—a written explanation of what SPC thought would happen in the next hour or two.

"I mean, it was just a classic. Rich Thompson was talking about what's going to happen," Wolter recalled. "And when he said, 'It looks like there's going to be a major outbreak throughout Oklahoma'—which is almost an exact quote from the movie *Twister*—it was just unbelievable."

Rich remembered it a little less dramatically and with a lot more frustration.

Rich and the SPC forecasters struggled with conflicting computer models. "If you added all the precipitation forecasts available from three or four different models, it could thunder from Kansas, Oklahoma, and Texas," Rich said. Every three hours, the computer kicked out numerical models. No two were in agreement. One of the models couldn't even agree with itself; it plotted a different result every time.

The frustrated forecasters finally threw out the computer models and worked from observational data generated by Doppler, the balloon soundings, and the satellites. At 1:00 p.m., the National Weather Service in Norman sent aloft an unscheduled balloon sounding. The grayish helium-filled balloon soared upward, its dangling miniature transmitter ticking back information to the ground-based computers. Currents of air pushed it in one direction and then another as it rose, and its seesaw flight confirmed the meteorologists' concerns. The winds aloft blew at different speeds: 20 mph from the southwest near the surface, 40 mph from due south at 3,000 feet, 35 mph from the southwest at 10,000 feet, and more than 50 mph from the west at 20,000 feet. The wind shear—different speeds at different heights—was definitely in play.

The midday sounding also showed the dryline cap—a warmer layer of air above the warm, moist air—was weakening and cooling in North Texas and southwestern Oklahoma. The cap acted as a lid on storm initiation. If the cap continued to weaken, the warm, moist air had a chance to break through, spiraling upward to form a storm.

A sheet of clouds over North Texas and southwestern Oklahoma also had thickened, adding to the uncertainty. The cloud cover could hold down ground temperatures, lessening the likelihood of warm thermals that would provide the lift needed by the warm, moist air to break through the cap.

"The joke is you could slam your car door and start a storm," said Rich. "It's not quite that easy, but these little subtle things are hard to observe. Sometimes it can be little terrain features. Or someplace that is irrigated farmland next to farmland that isn't irrigated."

Forecasters also measured the convective available potential energy (CAPE) based on balloon soundings. Using dew points at different altitudes and surface temperatures, they can estimate how fast a particle of air can rise. The higher the CAPE value, the faster the air particle soars upward. The CAPE gives meteorologists an idea of the strength of the warm updraft should it break through the cap. The CAPE values were highest near Wichita Falls, Texas, and into southwestern Oklahoma.

After the mesoscale discussion, a forecaster provided a plain English translation for Wolter and his students: there were probably going to be tornadoes somewhere nearby.

"My kids were just flying high," Wolter said.

They headed off to the University of Oklahoma campus to meet with Herb Stein, a member of the university research team that operated the DOW trucks. The DOW program was the brainchild of Josh Wurman, a hot-shot MIT grad with a penchant for engineering and a passion for weather.

Josh's revolutionary idea was to place miniature Doppler radars on flatbed trucks. The extended cabs of the trucks contained mobile, computerized laboratories where he could sit and observe the radar screens. This way, the trucks could keep pace with tornadoes and get close but not too close.

Stein was giving the students a tour of a DOW truck when someone asked if they were going out on the chase. "Naw, today's not looking too good," he responded.

Wolter debated what to say to such a premier researcher. "You know, we were just at the Storm Prediction Center, and they seemed pretty confident that something is going to brew up today."

"Really?" Stein responded.

"So he started calling on his cell phone. While he's talking on one cell phone, his other cell phone is ringing, plus he's trying to tell my kids about the truck. It was so great. His mind got so into going chasing, all of a sudden he's like, 'Uh, kids, yeah, this is a steering wheel.'

Stein got off the phone. "Wow, you're right. This is sounding really good. You guys should come along with us."

Twelve pairs of pleading eyes turned to Wolter.

"Mr. Wolter, why can't we go?"

"Because I don't want to be sued for the rest of my life. Besides, what would your parents say?"

Wolter also was accompanied by a student's father acting as chaperone. "Well, speaking as one parent, I trust these [DOW] guys."

Wolter paused. When he had proposed this field trip to the school board, he was asked if they were going to chase storms. He had danced around a definitive answer. He told them that Oklahoma was very flat and that tornadoes could be seen from great distances and the main reason for the trip was for the students to meet the nation's top meteorologists.

"Okay," Wolter told the students. "Let's go."

This instant the words flew from his mouth another, more cautionary thought entered his head: Holy Mother of Pearl.

5

THE TORNADO DETECTIVE

O N FOOT AND LEADING HIS YOUNG fiancée, Tetsuya Fujita followed the tracks of his quarry for six miles, past the broken homes and flattened rice crops and to the very edge of a green forest of tall, lean bamboo. His shirt hung limp on the unusually hot and humid day, but he pushed forward. He was trailing his first tornado. On September 26, 1948, a *Tatsu Maki*—Dragon Swirl—originated as a water spout and grew into a tornado as it moved through the bayside community of Enoura on Kyushu Island, a rare twister in Japan.

Tatsuko, his fiancée, whom he would marry a month later, joined him to "witness an act of God rarely experienced in Kyushu." The two were shocked to see the destruction to the community and the crops. They walked first to the inlet, found the landing point of the water-spout, and followed it inland as it became a tornado. He took note of the pieces of debris, how the roof peeled away from a home, the direction of a bent rice stalk, and would spend the first year of his marriage traveling from one weather station to the next so he could obtain the atmospheric and surface observations about this rare Dragon Swirl.

Eventually he produced sixty surface maps, each representing a twenty-minute interval of pressure, temperature, and rainfall during a twenty-hour period. It was his first tornado damage survey. The Dragon Swirl also prompted another idea. Fujita began to map the patterns of high winds and their effect on rice crops. How did the winds bend, break, or swirl the delicate plants? From observing the rice paddies, he could visualize the winds—their strength and their direction. It was as though he were studying a fingerprint.

On a cold day in January 1951, a letter postmarked from Chicago arrived at Fujita's physics office at the university. Fujita called it simply "the most important letter I received in my life." It was from Horace Byers, head of the University of Chicago's Meteorology Department and former head of the Thunderstorm Project. Byers enclosed a copy of the report from the Thunderstorm Project and praised Fujita "for noting the importance of the thunderstorm downdraft and outflowing cold air." It was the first of several letter exchanges between Byers and Fujita.

In 1953, Byers invited Fujita for a two-year visitation program at the University of Chicago. Fujita had the chance to work with the world's foremost expert in thunderstorms. Fujita booked a flight as soon as he could, which meant as soon as he could borrow $650 to pay for the Pan Am Clipper flight to San Francisco.

To Fujita, just about everything could be charted, graphed, diagrammed, or mapped. His relentless inquisitiveness rarely allowed for downtime, even when he was flying across the Pacific Ocean. Fujita spent the first airplane flight of his life sketching the clouds and plotting the time lines of the clouds. Two hours into the flight, the four-engine Clipper flew near a developing storm tower, and Fujita heard the sound of dishes and utensils crashing about the airplane kitchen and felt the bounce of the turbulence. "A moment later, the flight became smooth and I saw a beautiful arc of low clouds." His chart was a vertical cross section of his flight from Tokyo to Wake Island, Wake to Honolulu, and Honolulu to San Francisco depicting the range of observed clouds and the time. "Dashed red lines denote lack of observation due to sleep," he wrote.

Fujita was midway through his arduous journey when the Clipper descended through a blanket of gray clouds for a smooth landing at San Francisco's airport. As a Japanese national, Fujita was allowed to bring only twenty-two dollars into the United States. His first stop in

San Francisco, perhaps on counsel from friends, was a hotel owned by a Japanese family. He sought advice from the front desk manager on how to stretch the twenty-two dollars to cover his three-day train ride to Chicago. The manager handed him a twenty-dollar bill.

With his new twenty dollars, Fujita bought a three-day supply of Fig Newtons and Coca-Cola for the Chicago-bound train trip. As the train pulled away from the Southern Pacific Station in Oakland, he ate his first fig bar for lunch; one at breakfast as the train passed the Great Salt Lake; another for lunch as the train departed Green River, Wyoming; one for dinner during a stop in eastern Nebraska; and the last only for survival as the train rocked toward Chicago on August 12, 1953. He would never eat Fig Newtons again.

Without his wife and son, Kazuya, Fujita found himself lonely and his limited English a barrier. But there were no barriers in science, and he devoted himself to research projects.

During his first two-year stint at the university, Fujita joined with researcher Morris Tepper at the U.S. Weather Bureau in Washington, D.C., to study severe weather. They measured the hourly air pressure and temperature changes around two storm systems in Kansas and Oklahoma. The changes usually were so minute that forecasters routinely ignored them. Fujita and Tepper called the miniature low-pressure systems mesocyclones, a counterclockwise swirl beneath powerful storms. They suggested that the mesocyclone, the rotation within the thunderstorm, was a key factor in tornado production. Fujita, with his cartography skills, was able not only to explain but also to diagram the mesolows, mesohighs, and mesocyclones, the word *meso* indicating those weather systems up to 250 miles wide. At the same time, he and Tepper created a new language for researchers and provided forecasters with a critical clue for predicting a tornado. Their groundbreaking work was published as Weather Bureau Research Paper No. 39.

The very idea of mesoscale research was new. The violent weather on the plains rarely stemmed from the large-scale, or what meteorologists call synoptic, events, but from these smaller-pressure highs and lows that boiled upward as burly storms on spring afternoons.

STORM WARNING

The power and the behavior of the tornado fascinated Fujita.

He interrupted his Chicago research to briefly return to Japan to finish his teaching contract. In 1956, he returned with his wife and Kazuya. No fig bars this time, just a head full of ideas and a new name: Tetsuya became Theodore or Ted.

BEFORE FUJITA, TORNADO RESEARCH had moved little beyond the scope first established by John Park Finley seventy years earlier. The number and location of tornadoes were tallied by state, month, and year, and so were the deaths. Byers's Thunderstorm Project discovered much about the outlines of a thunderstorm but nothing about tornadoes. Fujita was about to change the way scientists researched, analyzed, and thought about the twister.

On the afternoon of June 20, 1957, motorists traveling to Fargo, North Dakota, reported to police that a "very large swirling cloud" was moving toward the city. The police issued warnings that were broadcast by radio and television. But instead of running for cover, scores of people ran outdoors with their new Instamatic cameras to snap its picture.

The large swirling cloud was not a tornado but the mesocyclone of a supercell. The supercell did spawn five tornadoes, the third plowing through Fargo, killing 10 people and injuring 103 others.

When Byers heard rumors of a large number of photographs of the storm and tornado, he asked Fujita to investigate. Rarely had a tornado and its parent been so well documented.

After appeals from a local television station, amateur photographers turned over more than 150 snapshots and home movie reels to Fujita. He began a painstaking two-year process that invented photogrammetry for tornado studies, detailed and named the cloud structure within the storm, and provided initial clues on the formation of tornadoes within mesocyclones. He visited the positions where the photographers had been standing to gauge the angles of the photographs. As he had with the cemetery flower pots in Nagasaki to compute the height of the atomic bomb, he triangulated the coordinates of the tornado from every photograph.

He reduced the photographs to a single size and recreated the tornado formation and touchdown through a sequence of shots. He was able to document the life cycle of the tornado and the mesocyclone from which it descended. Fujita referred to this lowering of the mesocyclone base as a wall cloud, and he identified the tail cloud and collar cloud. From home movies taken by residents, he was able to determine the speed of the rotation by measuring the speed of debris within the funnel.

Fujita's study was a landmark achievement, the first documentation of a supercell thunderstorm. It was, said his admirers, his seminal work. He provided terminology—*wall cloud, tail cloud,* and other terms—that would become the standard language of tornado research and forecasting. His innovative use of the film and photographs created a new way to study the phenomenon. No one had considered that all those formations within the storm cloud actually meant something. He concluded that "the tornadoes in the Fargo area did not occur by chance but were the product of well organized conditions very favorable for tornado formation." Detecting those conditions would preoccupy researchers.

By chance, researchers with the Illinois State Water Survey studying rain in powerful storms had discovered a "hook echo" effect on radar in 1953. Fujita and other researchers would associate that radar image with the swirling mesocyclone, giving weather forecasters at least a signature for potentially tornadic storms.

Fujita's observational method of research was one that he would use repeatedly. Where people saw only a tornado disaster, Fujita saw a clue to a mystery that, if studied in its details, would allow one to form a theory. If one could gather enough clues, the mystery of severe storms and their offspring could be solved—a meteorological crime scene investigation. Each tornado provided a tantalizing new detail for him to explore.

The Government Printing Office issued the report as U.S. Weather Bureau Research Paper No. 42, selling it for forty-five cents a copy. Most of the 3,000 copies of Research Paper No. 42 were purchased by the Fargo-area residents. They wanted to see their snapshots in print.

STORM WARNING

• • •

As A NEW DECADE approached, meteorology was about to take a quantum leap into the future. In 1957, the Soviet Union launched the satellite *Sputnik*. After trial and error, the United States responded with TIROS-1, a 270-pound weather satellite that housed two television cameras and was the first attempt to establish a worldwide meteorological satellite network. During its seventy-eight days in space, TIROS-1 provided thousands of pictures of Earth, large-scale cloud systems never before seen.

The weather satellite was operated by the National Aeronautics and Space Administration, which also was using it to test instruments for future space flights. Satellites became the meteorologist's eyes for large-scale weather systems. For Fujita, it was another tool with which to observe the clouds.

It was Fujita who put the satellite clouds in motion. Starting with TIROS and continuing through more advanced weather birds, Fujita showed meteorologists how cloud height could be estimated by looking at their shadows on the Earth and how wind speed could be estimated by cloud movement. He designed the "Fujita wheel," which allowed forecasters to flip through a collection of satellite photographs to see the actual motion of the clouds.

Applications Technology Satellite (ATS-1) was the first geostationary satellite, launched in 1966. Within months, Fujita made a movie showing mesoscale cloud patterns in motion. The film demonstrated to meteorology researchers and forecasters that animated satellite images could reveal atmospheric motion. For the first time, researchers and forecasters could see the air masses in action, enabling them to predict with greater certainty the weather potential from more minute cloud movements. This was especially helpful for viewing cloud movements over the oceans, where weather data were limited to a few buoys close to shore. Now armed with radar and satellite images, meteorologists had critical new tools for increasing the accuracy of weather forecasts.

Fujita studied the cloud tops just as he did the destruction on the ground. From a pressurized Lear jet, he documented the behavior of the very top of the cumulonimbus clouds, the way the warm updraft flew up, cooled, and sank. There appeared to be some correlations between the sinking of the clouds and tornado formation. For each cloud he studied, Fujita took scores of his own photographs and matched the cloud formations to satellite pictures from above and radar signatures on the ground.

The idea of using photography came from the Fargo tornadoes. Those smooth satellite loops now seen during every television weather broadcast began with Fujita and all those snapshots taken by curious Fargo residents.

"Ted was an idea man," said Charles Doswell III, a scientist at the National Severe Storms Lab. "Ideas flew from his brain like sparks from a grinding wheel."

CLOSER TO THE GROUND, other weather researchers had their eye on the Pentagon's early warning radar system at the Arctic Circle. Starting in the early 1960s, scientists began trying to convert the Doppler radar into a weather radar.

The U.S. Weather Bureau created the National Severe Storms Laboratory in 1964, and modifying the Doppler for use by forecasters became one of its primary missions. The lab was located in Norman, Oklahoma, with the idea that there would be enough tornadoes to conduct experiments.

The conventional World War II–era radar allowed forecasters and researchers to see only green blobs of clouds. The Doppler radar eventually would allow forecasters and researchers to see inside the clouds. The concept was developed by Christian Andreas Doppler based on sound waves. Doppler noticed that a train whistle becomes higher-pitched as it nears and lower-pitched as it passes by. The nineteenth-century Doppler effect was adapted to twentieth-century radar use. A Doppler radar could determine the intensity and direction of the winds and rain. Modified for thunderstorms, it could allow forecasters to see the telltale signs of rotation from the mesocyclone.

STORM WARNING

The Cold War proved to be a technological bonanza for the Weather Service. Before the 1960s, the weathermen—and they were almost all men—had used the same basic tools—thermometer, barometer, anemometer, and their own eyes—for centuries.

The scientists at the National Severe Storms Lab were on the frontier, literally and figuratively. From their Norman office, they initiated the first storm chases for scientific purposes; they studied storms for clues that would aid forecasters in their predictions, and they dragged their experimental radar across the plains. Slowly, these tornado cowboys on the prairie began to pick away at the twister's secrets or at least create new theories about the tornado. The primary goal of both researchers and forecasters was at least to increase the warning time to give the public a chance to seek shelter. And it was clear they had much work to do.

PALM SUNDAY 1965 WAS balmy throughout the Midwest as temperatures soared into the 70s. Many people were in church as Christians began marking their holiest week of the year. Those who did hear the weather report were confused. The Weather Bureau terminology for tornadoes was divided into *tornado forecast* or *tornado alert,* and little effort had been expended to educate the public on the differences. The weather in the Midwest had been chaotic for a week, unusually warm during the day and cool at night. At midday on Palm Sunday, an unusually forceful jet stream swept across the southern plains through Iowa and then eastward. A tornado alert was issued.

At 1:00 p.m., the first tornado formed in Clinton County, Iowa, and for the next twelve hours, twisters pounded the Midwest from Iowa to Ohio, Michigan to southern Indiana. Forty-seven tornadoes churned across six states, killing 271 people and injuring 1,500 others. Hardest hit was Indiana, where 130 people died and 1,200 were injured. Many of the dead were found in the wreckage of mobile homes after twisters marched through several trailer parks.

The next day, Fujita viewed the destruction from the cramped seat of a small Cessna. For four days, he flew 7,500 miles to map these killer tornadoes and confirmed a theory he had long suspected. Ever since John P. Finley's tornado investigations, people thought long-lived tornadoes traveled for hundreds of miles. Even the 1947 Woodward tornado was thought to be one lone twister that moved more than 200 miles through Texas and Oklahoma.

Fujita suspected storms actually produced a series of tornadoes along the same track. The thunderstorm cycled through a family of tornadoes, a theory he first suggested from the Fargo storm. And often, each time the storm cycled through a twister, it became bigger and stronger. The general assumption at the time was that tornadoes skipped along the landscape. Fujita's detailed map of the Palm Sunday outbreak showed that many of the tornadoes believed to be long-tracked were a series of brief tornadoes that grew stronger and more deadly each time the storm cycled through new funnels.

The Palm Sunday outbreak also prompted the Weather Bureau to reassess its tornado terminology. It scrapped the *forecast* and *alert* terms. Instead, it would issue *watches* and *warnings*. A tornado *watch* meant conditions were favorable for tornadoes; a *warning* meant a tornado was either imminent or on the ground, and it was time to seek shelter immediately.

While airborne over the Midwest, Fujita again was baffled by circular marks that, from the Cessna, appeared as though the tornado scratched deep into a field. Fujita had his own wild theory about these circular marks. He believed that there were smaller whirlwinds within tornadoes—that multiple vortices rotated within the bigger twisters. The circular marks from the Palm Sunday twisters would preoccupy him for a decade.

What was clear to Fujita and other researchers at the time was that not all tornadoes were created equal. Some were strong; some were weak. The Weather Bureau had been keeping count of the number of tornadoes but not their strength. The problem was how to measure the wind speed. Traditional anemometers collapsed during high winds, so actual measurements of wind speed were impossible.

One nascent sector, the nuclear power industry, needed to know the risks of violent tornadoes to help it build tornado-resistant facilities. The Nuclear Regulatory Commission and the National Science Foundation began providing grants to Fujita.

In 1970, Fujita traveled to Lubbock, Texas, where two tornadoes, including a violent twister, tore through town killing 26 people and injuring 1,500.

Five young engineering professors at Lubbock's Texas Tech University—Ernst Kiesling, Jim McDonald, Kishor Mehta, Joseph Minor, and Richard Peterson—surveyed the destruction and decided to see what, if anything, could be done to lessen the impact of these huge winds. The professors formed a new research center within Texas Tech, eventually calling it the Wind Science and Engineering (WISE) Research Center.

Fujita charted the damage path. He surmised that all but one person died when caught within the twister's more powerful "suction spot," where he believed the minitwisters within the tornado created even more power. He also met with the Texas Tech professors, the beginning of a lifelong association.

"He was brilliant and meticulous. He paid great attention to detail. I don't know if I've ever met anyone who worked any harder than he did," McDonald said. An informal partnership developed. After a tornado, Fujita could be found studying the destruction path from a high-winged Cessna. The Texas Tech engineers would walk the same path, studying the effects of the ground-level winds on structures.

Measuring tornado wind speed proved difficult. Fujita relied on the same method he used with the Fargo tornado, using films and photographs of twisters and estimating the speed based on a piece of visible debris and timing its movement within the frames of a film. Engineers, however, wanted to know the speed at the surface where the wind forces destroyed even strong buildings.

Fujita used the upper end of the twelve-step Beaufort Scale, the nineteenth-century system used to help sailors estimate wind speeds based on observations such as wave size, and the lower ends of the Mach scale, which estimates speed based on its relation to the speed of sound. The result was the Fujita Scale, a six-step range of wind speeds based on the damage caused by the winds and on the type of structure that was damaged.

In 1971, he unveiled his idea for measuring the twisters:

F 0 (40–72 mph, Light Damage): Some damage to chimneys, TV antennas; breaks twigs off trees; pushes over shallow-rooted trees

F 1 (73–112 mph, Moderate Damage): Peels surface off roofs; windows broken; light trailer homes pushed or overturned; some trees uprooted or snapped; moving automobiles pushed off the road

F 2 (113–157 mph, Considerable Damage): Roofs torn off frame houses leaving strong upright walls; weak buildings demolished; trailer homes destroyed; large trees snapped or uprooted; railroad boxcars pushed over; light object missiles generated; cars blown off highways

F 3 (158–206 mph, Severe Damage): Roofs and some walls torn off frame houses; some rural buildings completely demolished; trains overturned; steel-framed hangar-warehouse structures torn; cars lifted off the ground; most trees in a forest uprooted, snapped or leveled

F 4 (207–260 mph, Devastating Damage): Whole frame houses leveled, leaving piles of debris; steel structures badly damaged; trees debarked by small flying debris; cars and trains thrown some distances or rolled considerable distances; large missiles generated

F 5 (261–318 mph, Incredible Damage): Whole frame houses tossed off foundation; steel-reinforced concrete structures badly damaged; automobile-sized missiles generated; incredible phenomena can occur

Fujita also plotted his scale from F6 to F12 but deemed the destruction past F5 as inconceivable. Also, because the atmosphere has mass, it creates friction that ultimately would limit any wind speed. Fujita suggested surface winds could not exceed the F5 level, 318 miles per hour, which was his best guess. He also defined wind speed as the speed at the fastest quarter-mile. A wind speed of 225 miles per hour covers a quarter mile in four seconds.

The Fujita Scale became an imperfect guide for tornado measurements. An intense tornado rolling over an open field would receive a lower F-scale rating than if the same tornado struck a housing addition. And the Texas Tech engineers argued the wind speeds for violent twisters were too high—that even well-built homes could be destroyed at lower speeds. Still, something was better than nothing.

Fujita and Allen Pearson, then chief of the Storm Prediction Center, convinced the National Weather Service to begin assigning Fujita Scale numbers to tornadoes immediately.

Fujita also launched a massive effort to create a tornado climatology database dating back to 1916. The idea was to use news clippings and other documents indicating the damage caused by tornadoes and assign an F-scale rating. It would take nearly a decade to complete the task.

When it was done, scientists for the first time had an idea of both the frequency and intensity of tornadoes. The database provided researchers with some startling information: tornadoes with wind speeds above 158 mph, the F3s and higher, accounted for only 9 to 12 percent of all twisters but 88 percent of all fatalities.

The smaller tornadoes usually were produced by nonsupercell thunderstorms. The smaller twisters, which were far more numerous, were the most difficult to detect, but also shorter lived and less powerful, though far from harmless. The big beasts were the problem, and warnings alone would not save lives. People needed shelter from the storms.

6

PRIORITY ONE

A T MIDDAY ON MAY 3, GARY England stepped outside the KWTV broadcast center for the first time all day, stood in the circular driveway, and tilted his head toward the sky. A dozen storm trackers, a helicopter, the entire newsroom crew, and an anxious news director were waiting for him to make a decision.

Gary and his weather team colleagues had huddled in the television station's darkened weather center much of the day. They reviewed one numerical model after another. They studied the weather watches coming from the Storm Prediction Center and the National Weather Service office in Norman. They made calls to the scientists in Norman.

"We analyzed the data until we were cross-eyed," Gary said.

As with the meteorologists at the Storm Prediction Center and the National Weather Service, there was no agreement among Channel 9's weather team. The analysis ranged from a hard spring rain to damaging tornadoes.

Just two words from Gary—"priority one"—meant all of the station's reporters and camera crews would be at his disposal. It meant tearing up the evening newscast. It meant getting the station's helicopter, Ranger Nine, airborne. It meant deploying storm trackers. And all that cost the CBS affiliate money. There was the expense of the crews plus lost advertising dollars should the station start interrupting its broadcast with warnings.

Gary's priority system—one, two, or three, depending on the severity of the thunderstorm—determined the prominence of weather in the newscast. If Gary declared a priority one, something newsworthy better happen.

In the spring in Oklahoma, the weather was often the main news item of the day.

Gary's priority system was the product of many years of trial and error. But none of the meteorologists—coping with the mixed computer models, a confusing dryline, and the cloud cover to the south—at either the Oklahoma City station or the Norman forecast headquarters could reach a consensus. It was driving him crazy.

GARY ENGLAND MAY BE the most famous native of Oklahoma's Dewey County since prohibition advocate Carrie Nation emerged with her hatchet and a thing about whiskey. A sign at the city limits of Seiling, a small farm town about 100 miles northwest of Oklahoma City, proclaims itself "The Home of Gary England." There's also "Gary England Avenue."

A strawberry-blonde with a smattering of freckles, Gary grew up in Oklahoma's wheat country. He was born in the family's yellow farmhouse in the 1940s. He won't give the year. Short on cash, his dad paid the attending doctor with chickens and a handshake.

The 1947 Woodward tornado frightened and fascinated him as a boy. Relatives regaled him with tales of "cyclones" clawing across the Oklahoma landscape. He decided he wanted to be a television weatherman in the seventh grade and made weekly trips to his grandmother's to watch his hero: Harry Volkman, the weatherman on WKY-TV in Oklahoma City.

"On Sunday evenings, Harry had a fifteen-minute weather show," Gary recalled. "He talked about his family. He was just so personable. He did troughs and lofts and talked about the jet stream and upper winds. He just made it all so fascinating."

Gary wanted to be Harry Volkman, even when he wasn't sure what it was exactly that Harry did.

Volkman, a slender, gaunt-faced Army veteran at twenty-six, was one of the first TV weathermen in Oklahoma, first in Tulsa and later Oklahoma City. He appeared on Oklahoma City's WKY-TV every evening and every night, and on Sundays he had a program called "Weather Station." He wasn't a meteorologist, but, starting as a teenager, he had a special interest in weather.

He gathered his information from Weather Bureau teletypes and Morse code from short-wave radios. He used white chalk on a metal map of Oklahoma to indicate rain showers or sun. As he explained to *The Daily Oklahoman* in an interview, Oklahoma's weather was comparable to the moods of a woman: "Both are very changeable and you never can be absolutely sure what you are going to get."

He also had a program on WKY radio that bragged in an advertisement that Volkman "has unequalled facilities at WKY for his informative weather program—meteorology teletype machines which bring him official reports from almost every major city in the nation, every hour, plus reports from Canada and Central America and ships at sea."

Volkman, despite being wildly popular in the state, almost got canned. In the spring of 1952, he received information from Tinker Air Force Base's weather unit, the one operated by Miller and Fawbush, that a tornado appeared headed for Oklahoma City. He interrupted the regular broadcast to issue an on-air tornado warning. The WKY-TV station manager seethed; the U.S. Weather Bureau was furious. The Oklahoma City weather office already had a direct connection to WKY-TV that would allow it—not the TV weatherman—to issue alerts and, besides, no one had ever broadcast a tornado warning on television. "It was taboo," said Volkman.

The next morning, Volkman was on the verge of being fired when the station's telephones started ringing. Viewers were calling to thank him. His job was saved. And television assumed a new role in weather warnings.

Volkman, a Massachusetts native, stayed briefly in Oklahoma. In 1959 he moved to WGN-TV in Chicago, a bigger market, and remained on air for more than forty years. But he left a lasting mark on

Oklahoma television with his on-air tornado warnings. He also started a battle between television meteorologists and government meteorologists that Gary would carry on with glee.

Once the U.S. Weather Bureau decided to issue tornado forecasts in the 1950s, it still found itself with only the most rudimentary of weather tools. Its best tornado warning system came from its network of ham radio operators who functioned as the bureau's eyes on the sky during severe weather. The storm spotters alerted the local weather office when they spotted a developing tornado. The weather office would then call the local police or fire department and notify a local radio station. Often the police and fire would roll through town sounding their sirens on their cars and trucks. The system was not only inefficient but dangerous for the officers.

For a time, Oklahoma even considered a civil aviation patrol that would become airborne during stormy weather. The idea was to attach sirens to the planes and fly low over affected towns, sounding a tornado alarm. That flawed concept—airplanes and thunderstorms don't mix—never left the ground.

A more efficient warning system came courtesy of the Red Menace. Fear of a possible nuclear attack by the Soviet Union prompted the U.S. government to promote the creation of civil defense offices in states and counties around the nation. The government also offered to pay part of the costs for air raid sirens. Cities that could afford it took up the offer. In Tornado Alley, those air raid sirens quickly became alarms for tornado warnings.

The United States also upgraded its early warning system in case of a Soviet attack, deploying a new radar array around the Arctic Circle. This new Doppler radar system could determine the direction and speed of objects. But the U.S. Weather Bureau was stuck with World War II radar technology. It modified radar used by Navy aircraft to create the WSR-57 (Weather Surveillance Radar–1957), which it installed nationwide in the early 1960s. The WSR-57, with its viewing scopes and green masses of clutter and clouds, remained in service for three decades.

It was better than nothing, but at the weather office for Oklahoma

City, then located at the Will Rogers Airport, a storm had to practically be on top of the city before they could see it. Tall buildings, even grain elevators, obstructed its scans. The Doppler, if it could be adapted to weather uses, promised to be something special.

Gary planned to attend college after high school, but when he passed a Navy recruiting office one day, he had a better idea. Like many other meteorologists of his era, Gary got his training from the military. After a stint in the Navy and a degree from the University of Oklahoma, he became the weatherman for a local radio station. Gregarious and creative, he cooked up an imaginary sidekick, the thunder lizard, that kept him informed of the weather. They were a hit. KWTV came calling in 1972.

It was, he said, the Stone Age for weather forecasting. "There was no radar at this station. We had two chalkboard maps, one of the United States and one of Oklahoma. We had one teletype machine and one old facsimile machine that printed out Weather Service maps. The news director thought the teletype machine would look really good in the newsroom. During my first storm season, that teletype bell would go off and I'd run across the room and rip off that warning and run back out of breath and scaring the daylights out of people."

Gary quickly saw there had to be a better way than just relying on the Weather Service reports. Early in 1973, he charted central Oklahoma into one-mile squares and found ham-radio operators willing to work as storm lookouts. Each square had a number (Alpha 21, Alpha 22, and so on). "We set up a code so they could transmit back what they had seen and what direction it was moving."

Technology also was making it feasible for local television stations to enter into the forecasting business, which also helped their ratings and in turn increased advertising. KWTV bought its first World War II–vintage radar from Alabama-based Enterprise Electronics a year after Gary arrived. It beefed up its weather team, hiring two more meteorologists. The same year, the CBS network began providing its affiliates with a satellite loop once a day. The picture showed a jerky blur of clouds crossing the country, a successor to the concept developed by Fujita.

"We very quickly went from one teletype and one facsimile and the Weather Service warnings to having our own radar, our own spotters. But we were really flying by the seat of our pants."

The warnings remained terrible, Gary said.

"In those days, I could only warn Jim because down the road his neighbor Bob's house just blew away." Sometimes viewers would even call in urging him to issue a tornado warning because they could see a twister on the ground.

For Gary, the status quo was no longer acceptable. On June 8, 1974, a surprise tornado outbreak killed sixteen people in KWTV's viewing area. Twenty-two tornadoes struck the state that day, including five in Oklahoma City. Gary made a pledge to himself never to be surprised again, not on his watch. He began working more closely with the National Severe Storms Laboratory in Norman, which was working to modify the Doppler radar for weather.

In 1974, the Weather Bureau's lead time for a tornado warning was a negative ten to fifteen minutes. A twister had to be on the ground and confirmed visually by spotters before forecasters could send out a public alarm. The Doppler, with its ability to detect the swirling wind shears within the storm cloud, would allow forecasters to issue warnings a few minutes in advance. Not a great deal of time, but enough for a person to round up the kids and the pets and head for shelter. The Doppler would save lives.

"It was pretty obvious this was going to be the hottest thing to come down the pike, and it was going to stop some of these embarrassments on forecasting," Gary said.

In 1973, the National Severe Storm Laboratory's Doppler radar team captured the evolution of a tornado for the first time. A tornado touched down at Union City, just west of Oklahoma City. Researchers focused their Doppler prototype toward the storm. There was no real-time display, no instant data. They had to process the magnetic tapes later. It took months before they realized what they had: the entire life cycle of a tornado. They could see the intensity of the wind speeds within the storm and the tornado, the winds going in different directions. They could see the swirling reflections of the rain. For the first time, they could see the circular signature of the developing mesocyclone, the supercell's calling card. And they could see it in advance of a tornado. The implications for weather forecasting and weather research were enormous. If they could spot the developing mesocyclone, they could watch it for telltale signs of tornado development and give the public more time to seek shelter.

It was, said Don Burgess, a chief researcher and former college classmate of Gary, one of many "eureka" moments scientists had with the Doppler. For Burgess and all the lab researchers, it was an incredible accomplishment.

That same Union City tornado also got Gary in trouble with the news director. For the first time, though not for the last, he issued a tornado warning before the National Weather Service did. The news director and the National Weather Service were furious; only the Weather Service could issue a tornado warning. A note from KWTV owner John Griffin praising Gary's decision silenced the news director.

The Union City twister proved that meteorologists could see inside the bear's cage, and after the June 8, 1974, tragedy, Gary had to have one.

He convinced the Griffin family, which owned Channel 9, to invest $250,000, a handsome sum for the affiliate, into developing a smaller version of the Doppler radar. In a competitive market like Oklahoma City, the weather segment of the newscast played a critical role in attracting and retaining viewers. Each station touted its meteorologist's skill and accuracy in the commercial weather battles, and each spring season brought a new gadget to dazzle viewers.

But KWTV was about to up the ante. A news director, who opposed the idea, threatened to toss Gary from the station's 1,572-foot tower if it didn't work.

Gary asked Enterprise Electronics to build a commercial version of the Doppler system. "No one had ever heard of Doppler outside of the storm lab and a few of the research areas." It took a few years to make it all work. In 1982, Gary England became the first television weather forecaster—actually the first forecaster period—to use Doppler radar.

The first time Gary powered it up, he wasn't sure what would happen. "My rear end was puckered up around my neck." His only training was a couple of color pictures provided by the storm lab that showed the hook echo display that possibly foreshadowed a tornado. Still, the Doppler system was rather primitive. There was no computer to interpret the data. "You could see convergence, you could see divergence [the wind coming from and going into a thunderstorm], you could see the really strong winds, but you had to be just glued to it. You were suddenly swamped with data."

Using the Doppler, Gary regularly started issuing his own tornado warnings before the National Weather Service, which was still using the World War II radar. The NWS meteorologists were miffed and insisted only the Weather Service should issue tornado warnings.

"He didn't think we were very good, and he was right," recalled Ken Crawford, the NWS chief in Oklahoma at the time. Crawford got England on telephone and asked for his list of grievances and tried to address them. "We didn't get any credit," Ken said.

The low point in the relationship between KWTV and the NWS came when Gary, the news director, and a camera crew showed up at a meeting Crawford and his staff were having with a fire department official. The fire department, responsible for sounding the sirens in case of a twister, had decided to go with KWTV's unofficial tornado warnings. KWTV aired a piece about the meeting, embarrassing the NWS crew.

FOR THE NATIONAL WEATHER Service, the new name of the U.S. Weather Bureau, the 1980s would prove as disheartening and controversial as the 1950s.

John V. Byrne, President Reagan's NOAA administrator, made a shocking announcement: the nation's weather satellites, which cost $1.5 billion to launch, would be sold to the highest bidder, and the government would repurchase the weather data. He said 30 percent of the agency's workforce would be reviewed for termination. Some specialized forecasts, such as crop forecasts for farmers, would be discontinued.

NOAA also intended to review functions such as weather-monitoring stations, emergency weather radio broadcasts, and computer weather analysis for outsourcing to private companies. Byrne told reporters the administration had faith in the private sector and its ability to make the proposal work.

The proposal struck a nerve with members of Congress, consumer advocates, and citizens in general. "This is a grotesque giveaway. It is taking information that everyone has free access to and turning it over to a private monopoly to traffic and profit with it," complained Ralph Nader.

Ten months after it was proposed, Congress killed the plan to sell the satellites. The move to dismantle large sections of the weather agencies also lost steam.

Even as it got to keep its weather satellites, the weather agencies continued trudging through a budgetary and political thicket.

The National Weather Service entered the 1980s using the same equipment that it had in the 1950s. World War II–era radar was still in use. Keeping the radars up and running was a full-time chore. Some spare parts no longer were manufactured, and metal fatigue left many of them inoperable.

As late as 1978, forecasters at field offices still used clear acetate charts that they marked with air pressure, wind, and rainfall and overlaid these charts on a light table so they could see the "big picture" for their forecasts. It was the same method Miller and Fawbush used in 1948. They did have a bare-bones computer system but could not integrate all the data they received from the aging radars, their satellites, or even the storm forecasting center. Some of its electronic equipment operated on vacuum tubes. They still communicated with each other by teletype.

Clearly, it was past time for the National Weather Service to upgrade its equipment, but the agency also was coping with budget cuts. It was desperate to deploy the new Doppler radar. It also needed a new integrated computer system that would allow forecasters to have immediate access to all the tools at hand.

The National Severe Storms Lab (NSSL) already had a Doppler

prototype that it had been using for years. In 1979, it completed its formal report for the creation of NEXRAD (next-generation radar). There were two ways for the government to undertake such a project. It could allow bids for companies to replicate the prototype developed by storm lab researchers, or it could essentially let a contractor start from scratch using some specifications. Reagan policymakers within the Department of Commerce opted for the latter, longer route.

It would take nearly twenty years to get all the new radars in place.

Burgess, one of the weather Doppler's innovators, called the process painful and sad. The development of the WSR-88D (Weather Surveillance Radar–1988 Doppler) was out of the hands of NSSL scientists.

Because of the slow procurement process, it took four years just to allow the initial group of vendors to develop their own models. Another year-long delay came after one contractor protested the bidding of the project. Unisys, later bought out by Lockheed, eventually won the contract, but the work on the Doppler was slow and the costs began to soar.

Just as important was the Advanced Weather Interactive Processing System (AWIPS), the computer system that would pull all the forecasting tools together. Congress balked at that too.

In 1988, *The Washington Post* reported, "The weather service, once the world's premier meteorological forecasting organization, concedes it is now a second-class operation struggling to improve. The White House, which some say is still pressing an old ideological battle to 'privatize' the weather service, is refusing to invest more money in it, however."

The false alarm rate was 62 percent. Still, the Reagan administration tried to pull the plug on AWIPS, refusing to endorse its $15 million start-up cost. The Weather Service also needed an upgraded supercomputer to bring it up to at least European standards.

The weather agencies were pounded by the national media. The $2 billion modernization effort mushroomed to $3 billion. In 1991, *Time* referred to the National Weather Service as a "technological museum." By then, the Weather Service's order for five new satellites was stalled, and only one was orbiting the Earth; the government was in a snit with Unisys after the company bid $386 million to build the Doppler radars and two years later requested $250 million more to complete the job; it finally got a supercomputer upgrade, but AWIPS was behind schedule.

Time noted the Department of Commerce had "never shown much understanding or interest in the science or technology of weather prediction." But it also noted that the Weather Service "is one of the few government operations that give every American a tangible benefit for his tax dollar."

The National Weather Service eventually would get its modern tools, but at a cost. It reduced its number of local offices by 50 percent and cut the number of employees by 17 percent to free up budget money to pay for modernization.

The final tab for the upgrades was $4.5 billion. The Norman office received its Doppler in 1992, nearly a decade after Gary England enthralled KWTV viewers with his version. The last of the 116 Doppler radars was installed in 1998.

The procurement process was its own disaster and one that undoubtedly cost lives. Researchers later found that the number of tornado deaths declined by nearly half, to about eighty a year, after NEXRAD became operational.

NEXRAD allowed forecasters to increase the warning time and to greatly expand the number of tornadoes they detected. By 1999, the warning time was eleven minutes—not long but a remarkable improvement since 1974.

KEN CRAWFORD WAS ONE of those National Weather Service workers who took a bashing in the 1980s. His political bosses in Washington, D.C., wanted to dismantle the entire agency, but Crawford plodded along in quiet efficiency. He had worked for the Weather Service since he was a high school intern in his home town of Fort Worth, Texas. He was soft-spoken and deliberate, a slight hunch to his shoulders like a man who had spent his life bent over a keyboard and radar. He was a man of a great patience and ideas, one of which saved many lives on May 3.

It started with a 1984 Memorial Day weekend thunderstorm that dumped fifteen inches of rain on downtown Tulsa within a few hours. Flood warnings had been issued by NWS two hours ahead of time. The word failed to make it to the people as the squall stalled over the city just past midnight. High water swamped entire neighborhoods and downtown. Forty people from a mobile home park had to be rescued one by one by a life-flight helicopter. Flood damage cost $180 million. Fourteen people died.

"A wonderful warning went out, but nobody in Tulsa paid much attention to it. Key leaders in Tulsa simply didn't get it, or got it and ignored it. Those are the only two reasons that they could have not been aware of what was happening," Crawford said.

Green and hilly, Tulsa was founded in bottomland of the Arkansas River. Scores of its tributaries webbed the city. By 1984, Tulsa was the most flood-prone city in the nation, in addition to being in Tornado Alley. The 1984 flood finally spurred city officials to take action to mitigate future damage.

But it also started Ken thinking. The deadly flood was a horrible communication failure. How could county safety officials access the same real-time data as the National Weather Service?

Crawford envisioned a network of small weather towers across the state of Oklahoma. As a government employee, he could not undertake such a county-by-county observation system or raise funds for such a project. He enlisted agricultural engineers, meteorologists, and climatologists at the University of Oklahoma and Oklahoma State University, who had been working on a similar idea to help farmers determine air and soil temperatures.

The idea was relatively simple. They created a network of thirty-foot towers, each equipped with sensors to detect the temperature, rainfall, solar radiation, wind speed and direction, humidity level, and air pressure—the basic weather essentials, available every five minutes. The towers also could measure soil temperature and moisture levels every thirty minutes, a major help to farmers. A solar panel powered the equipment. The data traveled by two-way telecommunications, piggybacking on the same network used by the state's law enforcement agencies.

In 1991, the universities' researchers secured some temporary state funding, and the Oklahoma Mesonet was born. *Mesonet* referred to mesoscale weather system events that are about 250 miles wide. Still, Crawford was concerned that the money was temporary. The Oklahoma Mesonet needed grassroots supporters.

Even as the 116 tower sites, at least one in each of the state's seventy-seven counties, and sensor purchases were being made, the Oklahoma Mesonet supporters decided to build a fan club from the ground up. The group secured grants from the National Science Foundation and began training schoolteachers on its uses for the classroom. The first fifty teachers were given desktop computers and access to a toll-free 800 number that would allow them to dial up the Mesonet data. The Mesonet became a teaching tool on weather, environment, and agriculture.

"We trained fifty carefully chosen teachers from all across the state and made it clear to them we wanted them to become our disciples, that if we produce a good product, we wanted them to be willing to tell others we produced a good product."

More important, Crawford wanted to create tiny converts among the schoolchildren, especially among the children of state legislators. He was planning ahead. "Those kids would go home and tell their parents, as did the other children, what they were learning in school, and we would gradually build a following. And that philosophy began to work almost immediately. What we had hoped for happened."

Uses for the Mesonet grew far beyond a flood warning or a classroom gadget. Utility plant officials used the temperature data to produce electrical power more efficiently. Farmers, receiving reports from the wind gauges, determined the best time to spray crops. County emergency managers, measuring humidity and temperature, decided the best time for the public works departments to pour concrete or aided firefighters by reporting wind direction.

In one case, the Oklahoma Mesonet even solved a murder. Its data helped establish a victim's time of death and punctured the suspect's alibi.

"There were no guarantees, but we felt we had a better chance if people sharing the message were at the worker level: they either talked with senators' and representatives' kids, or they were civil defense directors or police department personnel—they likely went to high school with a senator or representative, and they could tell the story better than we. If the story was good, it would be believed; and if it was believed, then it stood a chance of being funded."

By 1994, the Oklahoma Mesonet and its kindergarten-through-twelfth-grade outreach program were fully operational and totally successful. The Mesonet was ready for the next step in Crawford's long-term plan.

With the demise of the Soviet Union and communism, the old county civil defense operations had become emergency management offices, prepared to respond to different types of hazards, especially weather. Mitigation, preparedness, response, and recovery became their mission.

Crawford wanted to give these emergency managers more control over the weather information and communications. Technology also was on their side. More and more computers became a part of government offices, and the Internet provided access to the World Wide Web. In 1995, the Oklahoma Mesonet launched OK-FIRST, a safety program providing training and software to county emergency managers. Crawford and his associates gave the county emergency managers a crash course in meteorology—a one-week course that stripped all the math and physics from the science and didn't take years to learn.

Like the teachers, the first fifty county emergency managers received desktop computers, again secured by funding grants for the Oklahoma Mesonet. Proprietary software also was added. They could access Doppler radar and data from the local office of the National Weather Service. But they also had access to the real-time data from the Mesonet towers, which gave them more exact information about what was happening just blocks from them.

OK-FIRST gave emergency managers the power to call their own shots—and the ability to save lives. Its first major test would come on May 3.

WITH THE WEATHER SERVICE mired in its own financial and technological woes, Gary England continued his own innovations.

Enterprise Electronics improved the commercial Doppler prototype, and in 1991, KWTV purchased the Doppler 9000XL, which included a computer that interpreted the reams of data. After using the computer for a while, Gary convinced KWTV to hire a software programmer to write a program that would estimate a storm's time of arrival for each city and town in the state. KWTV called the system Storm Tracker, modified it to fit all fifty states, and sold it around the country.

In an era of megamedia companies, KWTV remained family owned and more flexible in how it spent its money. David Griffin followed his father John as head of the company. They lived in Oklahoma. They knew about tornadoes. And they were willing to spend the money. KWTV remained committed to improving its own meteorological tools.

For years, even the simple act of displaying the counties under a storm watch or tornado threat had been difficult. In the past, Gary used an Exacto knife to cut a red-paper outline of every county in the state and posted it on a yellow state background so it would be visible to viewers. A KWTV employee wrote software that automatically posted an outline of the counties under a weather warning. The software display program also was sold to television stations nationwide.

In 1991, Val Castor, a Tulsa native, was a senior studying meteorology at the University of Oklahoma. Val had been storm-obsessed since he was a kid. He had been a washout as a Little League shortstop because he was often looking at clouds on the horizon instead of the batter. In college, he took up storm chasing as a hobby. He thought he was pretty good at it—good enough that he screwed up his courage one day and called KWTV.

"Hey, Gary, there's some kid from OU on the phone who thinks he can chase storms." Val overheard the receptionist and cringed. "The next voice I hear is Gary's. I'm talking to Gary England," he said, still awed. Gary told Val that none of the stations employed storm chasers, but he might be willing to loan Val a cell phone, one of those big Motorola bag phones, and provide him with radar data if Val turned over all his videotape to KWTV. "It didn't take a genius to figure out who was getting the better of that deal," Val laughed. "But I didn't care. Man, I was pumped."

Two weeks later, Val found himself chasing and videotaping a small tornado. Gary liked the tape and gave Val an autographed copy of his autobiography. Later in April, Val made a big score, chasing and taping three violent tornadoes, including one that was a mile wide. Gary was so pleased he gave Val a KWTV windbreaker. But even Gary conceded he could not keep giving Val trinkets.

He got him a job as a studio cameraman for the nightly newscasts. Eventually KWTV agreed to pay him as a storm chaser. Val was the first. By 1999, KWTV had twelve chase teams to supplement its regular news crews. Computer technology allowed the chasers to send digital photographs and snippets of video back to KWTV as they tracked the tornado.

"All through the years, we would say, 'How can we do this better?' We never once said, 'Let's create something great,'" Gary said. The proximity of the severe storm laboratory in Norman also has been a boon to the local television stations. He credits the scientists there with helping keep the broadcast meteorologists in Oklahoma City at the forefront of weather research.

The importance of the weather forecasting prompted a fierce competition among the three network affiliates in Oklahoma City. Gary had his own name for it: Weather Wars.

Throughout the 1990s, as each station deployed all the latest weather forecasting tools, they rushed to be the first to issue a warning, to get the video, to declare a tornado on the ground, even when there wasn't one. Some storm trackers were so inexperienced they didn't know the difference between a tornado and the scruffy cloud rags called scud that hang from a thunderstorm's back side.

It reached such a pitch that Gary once interrupted CBS's *60 Minutes* to say that a competitor who had just declared a tornado on the ground was wrong; there was no tornado. That lit up the switchboard. Even he admits that he went too far.

The whole thing had gotten out of hand.

"In a mad rush for better television ratings, an extreme escalation in television coverage exploded across Oklahoma," Gary wrote in his auto-biography, *Weathering the Storm.* "Severe weather reporting and storm tracking grew from ridiculous to irresponsible. It went beyond the limits of sanity and good judgment."

In fifty years, Oklahomans had gone from no warnings to too many. Four Weather Service Doppler radars monitored the skies above the state, not including the Dopplers owned by the television stations. The Weather Service had its Skywarn spotter teams, and each television station had its own storm trackers. No other state in the nation was as wired into the weather. The Oklahoma Mesonet gave every county emergency manager an immediate glimpse at the atmospheric havoc and enabled them to quickly sound the storm sirens. And anyone with a computer could instantly access information from the Storm Prediction Center and the National Weather Service office in Norman.

But by 1999, Oklahomans had been so bombarded by television storm warnings, TV gadgets, storm trackers, and insipid live coverage of nonevents that a sense of complacency had taken hold. They simply tuned it out. Except for the October 1998 outbreak, tornadoes had been scarce for the past five years. Weather warnings had become just more white noise. The Weather Service knew it; the television forecasters knew it.

CHANNEL 9'S ONE-STORY BRICK headquarters sits just north of Oklahoma City on what passes as a plateau on the prairie. The building sits alone on several clear acres, dwarfed by the giant television tower, balanced by thick metal guy wires that keep it steady in the high Oklahoma winds. Perched atop scaffolding, the distinctive Doppler radar, which looks like a giant soccer ball, sits off to the side. And it was here that Gary stood on the afternoon of May 3 staring at the sky and inhaling the air.

"The sky was chaotic. There was the thick, rich smell of moisture. The heat. The humidity. The air just felt unusual."

It reminded Gary of another day, June 8, 1974, the day of an unexpected outbreak. He stood in the parking lot silently staring up at zigzagging cirrus clouds and immediately dismissed all the confusing computer data, all the conflicting numerical models, all the fancy gadgets.

As he walked back into the station, he hoped he was wrong, but his gut and his experience told him he wasn't. He found the weather producer and uttered two words: "Priority One." He was in charge.

AT THE STORM PREDICTION Center, Rich Thompson wrote and rewrote the mesoscale discussion advisory. Each time he thought he was finished, he received new and more ominous information. "What I ended up with was not at all what I started out with."

At 3:23 p.m. he wrote: "Convergence on the dryline is not strong and a cirrus shield over the TX Pandhandle/NW TX/WRN OK should limit additional surface heating. . . . But visible/radar imagery has shown the first attempt at TCU over far NW TX as of 20Z within a break in the cirrus. Mid level flow and vertical shear will increase over NW TX and WRN Oklahoma through late afternoon . . . with an increasing threat of supercells near the dryline from 00–03Z. This area is being monitored for a possible tornado watch later this afternoon."

Translation: The cirrus clouds had parted ever so slightly over North Texas and southwestern Oklahoma. The first attempt at thunderstorm creation occurred just north of Wichita Falls, Texas, within that cloud break. Plumes of cumulus clouds, buoyed by warm thermals from the surface, flared but failed to initiate. And the speedy jet streak was nosing its way into the region. Supercells and tornadoes were now a real possibility.

At 3:45 p.m. the SPC staff held a daily shift-change meeting. As the forecasters huddled around for the discussion, a second storm tower erupted near Lawton, eighty-five miles southwest of Oklahoma City, where the cap was weakest and the CAPE the strongest. The National Weather Service issued a severe thunderstorm warning. This storm had some staying power; it broke through the lid with a fierce updraft, and almost immediately the mesocyclone began spinning. It went from a cloud puff to a supercell within minutes. Meteorologists in both the weather office and SPC hovered around computers, marveling at how quickly the storm gathered power.

Rich called Roger at home: "Wake up and get in here." The chase was on.

7

HIDING FROM THE BEAR

THE METEOROLOGICAL CLUES MOUNTED. THE ATMOSPHERE was gray and volatile. The National Weather Service tried to prepare itself, but its tools were aging and its technology lacking. Storm clouds appeared on radar as pea-green blobs, satellite pictures remained blurry, and the NWS's main mode of communication was the teletype. It was spring 1974. The previous year set a record for the number of tornadoes: 1,100. Scientists weren't sure if there were really more tornadoes or just more tornadoes being reported, but probably the latter. More people were moving onto the twisters' home turf. Suburbs sprawled across once-vacant fields and new homes filled the pasturelands outside of midwestern and southern cities.

On April 1, twenty tornadoes struck the Mississippi and Ohio valleys, killing three people. There was more to come. A low-pressure system east of the Rockies, near the Kansas border, pushed Gulf air over the nation's midsection. A polar jet stream dipped across the Midwest. By April 2, it was clear to the Storm Prediction Center that severe storms and tornadoes were likely for the next day. It issued severe weather watches for the nation's entire midsection, from the Canadian border to the Gulf coast, for April 3.

As the low-pressure system moved eastward, three giant squall lines, each with supercells embedded, formed ahead of it. The squalls marched one behind the other through the length of the Mississippi and Ohio valleys. In the early afternoon, the first tornadoes touched down in Tennessee and Georgia, and for the next seventeen hours, the twisters didn't stop.

The total: 148 tornadoes in 13 states, with 315 people dead and 5,500 injured. It was the largest tornado outbreak ever recorded: the Super Outbreak. The shear breadth of the destruction amazed meteorologists. The tornado tracks covered more than 2,500 miles. The Tri-State outbreak in 1925 spawned 7 tornadoes that traveled 437 miles and killed 746 people. The Palm Sunday outbreak produced 31 twisters that covered 831 miles and caused 256 deaths.

One of the hardest hit was Xenia, Ohio, located ten miles east of Dayton. Nearly half the town of 25,000 was wiped away by a broad F5 twister that killed 30 people. Many lives were saved because of the timing. Downtown businesses throughout the region routinely closed on Wednesday afternoons, and many children already had finished the school day. The Weather Service later estimated that so many schools in the area were destroyed that as many as 1,000 children could have died had the twisters struck earlier.

TED FUJITA AND HIS teams of student researchers were airborne early the next day, beginning the first of many aerial views of the devastated region. By the time of the Super Outbreak, Fujita was the best-known tornado researcher in the world. *National Geographic* dubbed him "Mr. Tornado."

His work had already provided meteorology with much of its language and many of its theories about tornadoes: the wall cloud, the tail cloud, the overshooting tops of clouds into the stratosphere, the mesocyclone, mesoscale studies, the cycling of tornadic supercells that formed one twister after another, and the Fujita Scale to measure the power of the twisters. And, as always, there were the carefully crafted charts and diagrams from the skilled hand of the cartographer.

By the 1970s, he also was divorced and remarried. He never shared his private life with his students. Now he was famous and people asked for his autograph. And he always signed "Ted Fujita, Mr. Tornado."

His ability to "see" the invisible—to imagine the unknown—lifted him far above the other researchers and meteorologists. His skills as an illustrator and cartographer were just as dramatic. He could explain the most complicated meteorological concept with breathtaking simplicity.

"The core of science is to 'see,' to 'visualize' the physical concepts that weave the observational evidence together into a coherent explanation of the phenomena responsible for those observations," explained Les Lemon, a former National Severe Storms Lab (NSSL) researcher.

"He had a rare gift to 'see' what most of us could not 'see.' Further, he also had the gift to illustrate these concepts in a unique fashion and with an artistic flair such that no one else could approach that ability. These illustrations, being so unique with such clear logic, were always easily recognizable as his, no matter the concept and no matter the context. I could always immediately recognize them as his."

Fujita learned to see the invisible from Professor Otsuka's silhouette figures in a college physics class. The Meiji College professor used these shadowy figures dancing across the classroom walls to illustrate concepts of physics. The idea was to see beyond the visible forces of nature and use the power of one's imagination. It was a concept Fujita grasped immediately.

"After his class, I used to imagine the invisible phenomena which are hidden behind the silhouette images." Motion, matter, energy, and force—the four cornerstones of physics—also provided the starting points for the study of thunderstorms and tornadoes.

Fujita's idea of multiple vortices, the idea of minitwisters embedded in larger tornadoes, was like one of those silhouette images. He could see the cyclonic marks scratched across the farmland, but he could not find the proof for his theory. It would take some corn stubble, a dust devil, and the Super Outbreak before he could make his case, and in doing so, he explained one of the most vexing and heartbreaking traits of the big wind: why one home is destroyed and the one next door spared; why an entire roof can be removed and the fine china remain untouched; why the tornado moved with such a random and vengeful hand.

During one aerial survey over an Illinois cornfield, Fujita again saw the looping scratches. This time he landed and walked over the cropland, only to find there were no marks. What he found were swaths of corn stalk stubble. There were no marks on the ground, only the illusion when seen from the air. It was actually circles of debris.

It was "as if some vacuum cleaner–like structure within the tornado was able to converge the debris into a pile but could not lift it off the surface." Fujita began calling these swaths suction spots. Some meteorologists thought his idea of suction vortices absurd.

It would take a chance sighting to make it all clear. Flying back to Chicago after a tornado survey, Fujita saw a giant dust devil. The dust made visible the darker multiple vortices within it. From this chance sighting, Fujita theorized that multiple vortices also rotated inside a tornado, creating these suction spots where winds were even more intense than the parent tornado. These suction vortices, churning within the tornado like the horses on a merry-go-round, made sections of a tornado far more powerful than other parts of the twister.

To help prove his point, Fujita designed a tornado simulator that, using dry ice, could create small, visible vortices. Fujita's tornado simulator made a great visual for the media, and he often entertained reporters and visitors. But still there was no proof for his theory.

Once again, as in Fargo, it would be curious citizens who supplied the evidence. Photographs taken during the Super Outbreak clearly showed numerous tornadoes with vortices twined in their core before the dirt and debris obscured the view. "Some meteorologists did not believe my story until vortex pictures became available," he noted.

Once he latched onto an idea, he didn't stop. He was relentless. His graduate students at the University of Chicago occasionally hid from him so they could work on their own doctoral projects. The Super Outbreak provided him with a new challenge: to map every mile of every tornado.

There was little small talk after Fujita became airborne: the engine noise made it difficult, and Fujita became too absorbed in his work. The speedy Cessna 310 flew high and then spiraled downward as Fujita took hundreds of photographs.

He divided his graduate students into three teams to travel in Cessna airplanes. Scientists from NSSL and the University of Oklahoma also joined. Fujita's student, Greg Forbes, remembered flying in the high-winged Cessna. The pilot would fly low and level so Forbes could map the path. But the Cessna had to bank and circle the damage so Forbes and the students could take pictures. They spent days circling at a 45-degree angle. The expedition took not only a great eye but a strong stomach.

When one student in Forbes's plane became ill, the pilot set the Cessna down on a grass strip in Indiana. However, they soon saw they were on a short runway with a barn and power lines at the end. The pilot worried about clearing the obstacles with the weight of four people aboard.

"We all decided that we'd rather go for it than have to call Dr. Fujita to tell him we had to call a tow truck for our Cessna. We cleared the barn and power lines by about 3 feet as I saw it," Forbes recalled. Besides, Fujita paid for the planes personally, confident of repayment from a future grant.

For the students, crashing was a better alternative than telling Fujita they had failed. Fujita demanded total concentration. No clue, no seemingly innocuous corn stalk or bent tree, should be overlooked. The wind provided the answers; the scientists needed only to see it.

For the Super Outbreak, Fujita also produced one of his classic diagrams: a stunningly detailed map of each damage path from each of the 148 tornadoes, each with an F-scale rating. The Super Outbreak map also helped secure the Fujita Scale as *the* tornado measurement.

The Fujita Scale was imperfect. The government's Storm Prediction Center began its own historical database to assign F-scale measurements. Because of the subjectivity involved, the government information and Fujita's own database contained some major differences. A young Tom Grazulis, working for the Nuclear Regulatory Commission, which funded creation of the F-scale databases for both the Storm Prediction Center and Fujita, was assigned to ask Fujita to change the ratings.

But over glasses of vintage sake, Fujita talked and Grazulis listened. He didn't bring it up; he didn't have time. "Ted was describing and showing photographs of how he had used a hot steam iron and a block of dry ice to duplicate some usual vapor formation and temperature contours on and around overshooting thunderstorm tops." He would later recognize the hot iron conversation in Fujita's detailed report on storm cloud analysis, the result of his Lear jet excursions.

"For the next decade, at every meeting, I would see him pour forth idea after idea, most of them unpublished, all of them explanations of complex, three-dimensional meteorological puzzles. They would bring together radar data, damage surveys, satellite photography and analogies from other parts of the physical world and bring them together with such ease and simplicity that I felt stupid for not thinking of it myself," Grazulis wrote.

Not all of Fujita's ideas were brilliant. "He generated some bad ideas and he would defend them as vigorously as his good ideas," recalled Charles Doswell III, a storm researcher with NSSL. "He was kind of reluctant to admit that he ever had a bad idea. Frankly, the fact that he had an ego doesn't bother me. You have to have an ego to be a scientist. It doesn't mean that it somehow made him a lesser man; I think it made him more real to me."

Fujita himself once told his graduate students, "Even if I am wrong 50 percent of the time, that would still be a tremendous contribution to meteorology."

During his aerial surveys of the Super Outbreak, Fujita saw a strange starburst pattern. In an Appalachian forest, giant trees were felled in all directions, 360 degrees. It reminded him of the starburst patterns he saw after the atomic bomb explosions over Hiroshima and Nagasaki. Once again a silhouette danced before him. An invisible force was at work, and it would take him on another quest, one that would, once again, put his reputation on the line and close the circle of his life's work.

THE SUPER OUTBREAK SHOCKED the National Weather Service. Its meteorologists in thirteen states had issued 150 tornado warnings.

The only communication tool was the teletype. Warnings had to be keypunched by hand into a paper tape and fed through a reader. The warnings came so frequently in some states that the media wire services could not relay them fast enough. The weather office in Columbus, Ohio, had no way to contact city emergency officials, not even when their vintage radar saw the hook echo outlines of a tornado nearing the city.

Warnings were made, but in many cases people never heard them.

Texas Tech's Ernst Kiesling recalled seeing a photograph of the destruction at Xenia, Ohio. Amid the rubble of a flattened house was one small room still standing: the bathroom. The Texas engineers began studying the idea of in-home shelters. Smaller rooms such as bathrooms or closets better withstand tornadic winds because the interior wall studs are closer together, providing extra reinforcement. Interior walls within the center of a home also are the last to collapse in extremely strong winds.

After the 1974 Super Outbreak, followed in June by a smaller outbreak in Oklahoma, it was clear the National Weather Service needed to do more to educate the public on tornado preparedness. It expanded the number of stations for NOAA Weather Radio. Because of the large number of public schools destroyed on April 3, it began an education program for students and administrators. Hallways and small interior rooms were the safest places, gymnasiums and classrooms the most dangerous.

At home, if basements were not available—and most homes in Texas and Oklahoma did not have basements—people should seek shelter in the center of their house, in a bathroom or closet, and cover themselves with a blanket or mattress. It wasn't the winds that were so deadly; it was what was in the winds. The gravel, lumber, trees, glass, and cars all became deadly shrapnel in a violent tornado. The same shards that peeled the bark off trees could tear a human apart.

Compared with the excitement and daring of storm chasing, Texas Tech's work was decidedly unsexy. There would be no blockbuster movies or IMAX specials about construction codes. The Wind Science and Engineering (WISE) Research Center grew into a 56,000-square-foot indoor laboratory. Engineers created a wind cannon, a pneumatic contraption that when loaded with a two-by-four could hurl lumber about 100 miles per hour. Lumber pierced regular home exteriors, even brick exteriors. It provided a visual clue about the vulnerability of the average home owner.

No one had really studied the implication of impact and debris or how to mitigate the forces of wind. There were really only myths about the winds of a tornado. When tornado studies began, people assumed a twister could reach the speed of sound. No one knew how powerful a tornado could be.

For years, the National Weather Service had urged home owners to open the windows of their home before a tornado approached. The thought was that different pressure levels caused the house to explode. Texas Tech recommended that NWS drop the tornado tip. Opening windows only delayed home owners in seeking shelter. Besides, debris from the tornado opened the windows anyway.

"We got the notion that if we focused our efforts on a small interior room and hardened it, it could economically be made into a reliable shelter. It's not economical to do the whole house, but it is economical to take a small room such as a closet and bathroom and stiffen it," Texas Tech's Kiesling recalled.

Texas Tech published its first study in 1974, after the Super Outbreak, but the idea of an in-home shelter did not catch on. The WISE center also worked to encourage the coastal areas to strengthen construction codes to ward off the damaging winds from hurricanes. It had less luck in Tornado Alley.

As a tornado is whipping a home like an egg beater, it's the strength of the construction that could determine whether a family lives or dies.

Generally, construction codes in most U.S. cities call for houses to be built to withstand winds of 90 miles per hour over a three-second span. The weight-bearing "load path" for homes is based on gravity. The roof is attached to the wall frames, which are attached to the foundation. Frame-built homes are not constructed to withstand the side and upward pressure from high winds.

STORM WARNING

A Texas Tech University report explained how a two-story home disintegrates:

> *Visible damage generally initiates at the roof with a loss of a small percentage (less than 20 percent) of roofing material. Windows and door glass begin to break from flying debris. This action is followed by removal of additional roofing material and uplift of part or all of the roof deck. Garage doors collapse inward or outward, depending on wind direction.*
>
> *Internal pressure develops as a result of the broken windows or failed doors. Part of or all of the roof structure lifts up and is carried away by the winds. With removal of all or part of the roof structure, walls are no longer supported at the top. Exterior walls of the top floor collapse first, followed by the interior walls.*
>
> *As damage progresses, the second-floor floor structure is lifted up and removed. This leaves the first-floor walls unsupported. Again, the exterior walls collapse first, followed by destruction of most first-floor interior walls except possibly at small rooms, hallways or closets. The last degree of damage represents total devastation of the two-story residence. In a very intense tornado, this sequence of events takes place very rapidly. The roof and walls break up creating flying debris that adds to the destruction.*

This TTU sentence seemed a bit understated: *In a very intense tornado, this sequence of events takes place very rapidly. Rapidly* meant twenty to twenty-five seconds.

For a mobile home, even weak tornadoes cause havoc. Properly anchored, mobile homes can withstand a 70-mile-per-hour wind. "Above this range, a manufactured home will experience some form of damage," states the Oklahoma Manufactured Housing Association. "Only in the case of severe weather, such as a tornado, are these areas likely to experience winds in excess of 70 miles per hour."

In hurricane-prone areas, mobile homes are required to meet higher standards—100- to 110-mile-per-hour winds depending on location. High winds can push a mobile home off its moorings, rolling it like driftwood. So associated with tornado damage are mobile homes that the Manufactured Housing Association of Oklahoma created a "Myths and Reality" page that addresses one myth: manufactured homes seem to attract tornadoes. "There is no meteorological or scientific basis to thinking that manufactured homes attract tornadoes. The reality is one of coincidence: most manufactured homes are located in rural and suburban locations, where meteorological conditions favor the creation of tornadoes."

In tornado-prone locations such as Oklahoma, mobile homes are an affordable and viable housing option, a way for a family of modest means to get a toehold on the dream of home ownership. And it's doubtful that anyone living in a mobile home in Tornado Alley is unaware of the dangers during severe storms. Despite the threat, Oklahomans, like many other people, often feel "it can't happen to me." They wait until it is too late to seek appropriate shelter.

The coastal areas, especially Florida, have been more accepting of wind mitigation efforts suggested by Texas Tech. "In hurricane-prone regions, there has been more attention paid to standards and codes and enforcement has been better," Kiesling said. "In tornado regions, there is a somewhat flippant attitude because the incidence of tornadoes and the probability of a tornado occurring in a given location is pretty small and a relatively small percentage of the homes are affected."

But even in Florida, it took Hurricane Andrew to force cities and counties to enforce existing construction codes. There were numerous reports of shoddy construction practices and the lack of inspection and enforcement over home construction. The state in 2002 enacted a uniform, statewide construction code intended to mitigate hurricane damage. Miami-Dade and Broward counties passed even tougher rules.

Hurricane Andrew, with winds in excess of 175 miles per hour, killed 25 people and caused $25 billion (in 1992 dollars) in damages. More than 135,000 single-family homes and mobile homes were destroyed or damaged. Until 2005's Katrina, Andrew was the most expensive hurricane in history. There were some estimates that if the tougher construction codes had been in effect in 1992, the damage caused by Andrew could have been $10 billion less.

The new Florida code requires buildings in parts of the state to withstand winds of 110 miles per hour, even 150 miles per hour in the Keys. Along the coast, glass windows are required to be laminated to reduce the slashing shards, and hurricane straps are required to hold the rafters to the wall frames.

Texas Tech and its allies proved there are steps people can take to make themselves safer from the winds, whether from a hurricane or a tornado. It also studied the idea of reinforced in-home shelters. The Federal Emergency Management Agency, working with Texas Tech, produced a brochure in 1998 touting the benefits of constructing an in-home shelter and setting shelter standards. But there was no public stampede to comply.

Kiesling and his Texas Tech colleagues knew that upgraded building codes and in-home shelters could make a huge difference in Tornado Alley, but getting the public's attention and support was another issue altogether. It would take some monstrous event to call attention to shelters, just as Andrew had with Florida's building codes.

THE NATIONAL WEATHER SERVICE marked the twenty-fifth anniversary of the Super Outbreak on April 3, 1999, with a special presentation at Xenia, Ohio. It was a chance for the Weather Service to showcase its $4.5 billion modernization effort. All the new Doppler radars were in place; the new computer program, AWIPS, was up and running; new satellites not only displayed the clouds with greater clarity but provided infrared pictures of warm versus cool clouds.

The Weather Service issued a prescient news release: "Deadly storms such as the 1974 super outbreak can and will happen again. On average, across the country, the National Weather Service has doubled its warning lead times for tornadoes. But these warnings mean nothing if people don't receive them or don't take appropriate action after receiving them."

It would take exactly one month for the statements to come true.

8

INSIDE THE BEAR'S CAGE

A METEOROLOGICAL ARMY LAY IN WAIT NEAR Lawton, in the southwestern corner of Oklahoma. Two of Josh Wurman's mobile Doppler trucks were there, with the two vans full of Minnesota high school students parked behind them. Erik Rasmussen's VORTEX-99 mobile Mesonets were nearby, manned mainly by meteorology student volunteers. Several Oklahoma City television storm trackers sat in idling SUVs that brimmed with tiny microwaves to beam pictures back to the stations, onboard computers, cameras, and cellular phones. And there were a dozen more just plain curious storm chasers who were following the caravans or who on their own had pinpointed the most likely location for the thunderstorms to begin.

The crowd waited at midafternoon not for rain but for sunshine. If there were going to be thunderstorms, the high cirrus clouds would first have to break apart. The clouds would have to step aside as the dryline made its move eastward and the jet streak huffed into the region, which allowed the sun to toast the prairie. That was how it began. The same thermals that allowed hawks to glide silently in search of prey would provide the flinty spark to create a storm. The warm air thermals created the lift, the last of the necessary ingredients—the others being moisture, wind shear, and instability—to fall into place. And it was here, around Lawton, that it was the warmest and most humid and where the atmospheric energy—the connective available potential energy (CAPE)—was the greatest.

The clouds began to splinter over North Texas, and the sun began to shine from near Wichita Falls and into Oklahoma. A vertical cloud tower shot straight into the sky, but nothing happened. The engine failed to catch.

Val Castor and his fiancée, Amy Johnson, were Gary England's premier storm tracking team. Val was News 9's go-to man in the field; he was calm, articulate, and also a meteorologist. Val heard the radio traffic from other storm trackers. The North Texas storm fizzled. The sunny gap in the flimsy cirrus drifted directly over them. "We're in the perfect place," Val whispered to Amy.

Val watched Doppler radar scans displayed on his laptop computer that rested in the custom-made console of his black Silverado. Between them, there were two cell phones, a two-way radio, and a Global Positioning System (GPS). Affixed to the Chevy's roof was a small microwave transmitter. The News 9 Storm Tracker logo painted on the Chevy's doors let everyone know where Val was, which was good for newcomers. For the inexperienced chaser, it was easier to track Val than to track a tornado.

Amy operated the Sony video camera attached to a pole mounted in the floor on the passenger side. Once the action started, her job was to film the tornadoes and try not to get annoyed as Val said, "You are recording, aren't you?" All of Val's chase partners rolled their eyes at that question, which he repeated several times during a chase out of habit, as if they were there on a Sunday drive. With the computer, cell phones, and microwave, they could send stills or short eight-second videos to KWTV for broadcast.

There was no sense of alarm, just the excitement and anticipation of waiting for the atmosphere to give birth. "It was like a hundred other days in Oklahoma," Val said.

At 4:15 p.m., the National Weather Service office in Norman issued its first severe thunderstorm warning of the day. The air currents repeated the same dance they had made earlier over Texas. Val made up a game plan in his head. Once the show started, there was no time to think: he had to drive, keep one eye on the tornado and another on the road, figure out which road to take to keep pace, and give a blow-by-blow account to News 9's half-million viewers. One bad road decision, and the tornado would leave him behind. He planned to stay thirty miles or so upwind of the cloud gap. The distance gave the storm time to develop and gave him some space to stay ahead of it should it turn tornadic.

Once again, a little cloud puff grew visible as the warm air soared upward, carrying the icy drops higher and higher. A cloud poof sprouted like white, fluffy popcorn that fattened by the second and quickly grew dark, heavy with moisture. Nature cranked some more, and the cumulus tower spiraled upward: 10,000 feet and then 20,000; latent heat and moisture poured into the warm updraft, and it punched through the cap. No false alarm this time. The cumulus turned into a cumulonimbus, the thunderstorm. Downdrafts formed in the tower, creating gust fronts in the front and back. At 30,000 to 40,000 feet, cooler upper-level winds pushed and shoved the tower top, sprawling it across the sky. The cumulonimbus inhaled some more. At the ground, the surface winds picked up speed and hurried toward the thunderhead; the mushrooming storm banged against the stratosphere and flattened into a distinctive anvil shape.

The storm chasers watched in awe as the thunderhead flew skyward. The updraft hummed along so quickly that slower air around it began to spin from the updraft force. The entire base of this cumulonimbus rotated horizontally, a mesocyclone. Nature's perfect little engine crackled with power.

Back in the KWTV studio, Gary interrupted regular programming. Camera three swung around for a live shot of him from the weather center, the meteorologists' cubbyhole next to the newsroom. "We're in the beeps," a news producer warned Gary of the audio beeps that were interrupting News 9's broadcast. A "FIRST WEATHER" graphic highlighted in red was placed in the corner of the TV screen. A color map of southwestern Oklahoma showed a yellow and green mass just north of

Lawton. There was a second, small mass farther to the west near Altus. The camera flashed a red light. Gary was live.

"Conditions are pretty favorable for some explosive thunderstorm development. In fact, some of that is going on right now. This main storm is now developing very quickly," Gary told viewers. He pointed to the burgeoning storm clouds on the map in southwestern Oklahoma. He urged people to stay alert. He closed with his signature line: "We'll keep you advised."

Off camera, News 9 was crackling like a wartime submarine with cramped quarters and all hands at battle stations, with Gary as commander. Tiny magnets had the name of each chase team taped to them and were affixed to a metal state map. The magnets moved to pre-positioned spots. A staffer kept an open telephone line to Leroy Tatum, pilot of the station's helicopter, Ranger 9. A producer deployed news camera crews across the southern reaches of the city and told them to stand by.

The chase armada outside Lawton began to move as the warm updraft and cool downdraft continued their battle inside the supercell. The downdraft pulled down ice particles, and the updraft pushed them back up, with the particles bobbling up and down, growing bigger with each trip. A powerful updraft created hail the size of baseballs. What the updraft did not shove back into the storm began to fall as rain, heavy and thick. The chasers watched as the first storm, Storm A, organized itself by its atmospheric playbook: first came the rain, then the hail, and then the dark, ominous mesocyclone bringing up the rear.

At the National Weather Service office, the Doppler radar image of Storm A resembled a tadpole with its heavy rain and hail bulging out front and the sliver of its tail, the mesocyclone, curled into the distinctive hook echo. The Doppler reflected the storm as bright red and green—the high winds and rain swirling within the hook—moving toward (red) and away (green) from the radar.

At 4:45 p.m. the Storm Prediction Center (SPC) issued its first tornado watch, meaning conditions were favorable for "tornadoes, hail to 3 inches and wind gusts to 80 mph." Storm A's mesocyclone belly began to lower closer to the ground, maybe only a thousand feet from the surface. Near the mesocyclone, the wall cloud began to hang even closer to the ground. It was, as Fujita had found decades earlier, another ominous sign. It was like the opening of a bomber's bay doors.

Two minutes after the SPC watch, the Weather Service issued its first tornado warning. At 4:51 p.m. Storm A produced its first twister. A thin funnel barely twenty-five yards wide touched down for less than a minute at the northern edge of Fort Sill, the U.S. Army's artillery training ground north of Lawton. "Gary, there's a small one on the ground," Val reported to Gary. "Whoa, there goes something. It's beautiful."

The DOW trucks positioned themselves on either side, capturing the tornado as planned. The Minnesota kids piled out of their vans as Josh gave them a peek at the computer screen. The VORTEX cars followed along, the anemometers whirling atop them. It would be the only time the DOW trucks could position themselves on either side of a twister.

A few seconds later, the supercell produced a second, equally weak tornado. Moving slowly, it began to heave and grow as it suctioned more power into its engine. KWTV interrupted its afternoon shows, and Gary began broadcasting live, with Val's audio carried from a cell phone and video from News 9's helicopter.

"Val, tell me what you see."

"Gary, this thing has lost its identity for the moment, but it looks like it's trying to cycle back up again. Right now, I can see a lowering starting to form again."

"Val, we're looking at Ranger 9 [video] right now. We see a very broad wall cloud, a lowering, and that would be a little to the northeast of Lake Ellsworth."

"It looks like to me, Gary, this thing is rapidly getting organized. Its rotation has gotten much stronger."

"Val, it's a huge, huge lowering. The intensities on the storm are increasing."

"This whole updraft base is huge. It's a lot more ominous looking than it was over Lawton."

Only minutes after Storm A formed, another supercell flew skyward to its northwest. Storms started exploding across the prairie, each with such a powerful updraft that it turned into a supercell. Storm A was the most menacing, its twisters becoming more and more powerful. Storm A's third funnel, a 100-yard-wide twister, dug furiously into the ground near Apache, seventy miles southwest of Oklahoma City. The fourth and fifth tornadoes were brief and weak, almost like afterthoughts. But the sixth would have more staying power.

News 9's studio cameras were now trained on Gary for the duration. A map displayed the areas under tornado threat. "Val, there are storms going up everywhere. Talk to us about the storm you're on."

"The wall cloud is getting bigger. The rotation has increased dramatically in the last sixty seconds. The lowering has gotten a lot lower. There it is!" shouted Val. "Tornado on the ground! A brief touchdown!"

The twister, later numbered A6, formed just outside the town of Laverty, fifty-five miles southwest of Oklahoma City.

"I see it. We've got dirt and debris on the ground. Tornado on the ground. You're the meteorologist, talk to me."

"Gary, this thing is getting bigger. It's a big circulation on the ground. It looks like it could be multiple vortex to me from here. From the rotation, this is going to be a big one I think."

Val drove and did the play-by-play, the audio carried live over News 9 and simulcast on the radio, which played on the Meatwagon's AM receiver as Roger and Rich tried to catch up with the storm. Amy taped the tornado as it circled across pastureland and, using a computer and cell phone, quickly transmitted a still photograph back to the News 9 studio.

Just as Val predicted, A6 began to expand until its ground-level girth was nearly 900 yards.

Gary warned southwestern Oklahoma viewers again. "There's a large multiple-vortex tornado. It's very dangerous. Take shelter immediately." From his monitor of Ranger 9's camera shot, Gary saw bright explosions as the twister's tail struck power lines.

"Gary, it's hitting something right now. There's a lot of debris in the air. Lots of debris. This thing is just tearing things up! It's very intense at the surface. Very intense," Val reported. Val was driving a half-mile behind the twister and the Silverado rocked from 60-mile-per-hour gusts. "I don't see anything in the direct path of it, but I wouldn't want to be in the path of it."

AFTER TOSSING THE FOOTBALL with Jordan, Kara made him his favorite meal: macaroni and cheese. Or maybe she was just putting it in the microwave. The memory had dimmed over the years. But at that moment, there was no sense of urgency, no panic.

Her brother called. "Are you watching the weather?" Dustin asked. Kara and her mother spoke every day, but it was only recently that Dustin had started to call her out of the blue. "We had become more like friends," Dustin explained. She assured him she would keep an eye on it.

Her mother called minutes later. She had been planning to drive to Bridge Creek to watch her grandson play T-ball. "I asked her if Jordan's game had been cancelled. She said yeah, actually she said, 'Yes, mother.' " Mary laughed remembering Kara's exasperated inflection.

KWTV's HELICOPTER WAS THREE miles away from the main core, but the warm updraft rushing into the storm was so strong it tugged Ranger 9 toward the maelstrom. Ranger 9 pilot Leroy Tatum aimed the remote camera attached to the belly of the craft toward the dark cylindrical form and struggled to keep the chopper steady. A second twister appeared: a pencil-shaped satellite. Roger and Rich had parked the Meatwagon a safe distance away and stared at the twin twisters in amazement. When Gary England saw the dual tornadoes on a camera monitor, he—like Roger and Rich—knew that this was a day like no other. "Jiminy Christmas," he said aloud.

"This tornado is due west of Chickasha. It looks like it's beginning to crank up again. Val, keep talking to me about what you see."

"The updraft looks like a big flying saucer. It's just huge, and it's rotating fast."

Like Roger and Rich, who watched roadside, and Val, who chased a half-mile behind, Gary realized the twin tornadoes—the sixth and seventh formed by Storm A—indicated enormous energy within the storm. Not even Gary had ever seen such a sight as the twins. His Doppler radar screen indicated the supercell's mesocyclone region, the hook in the hook echo, was six miles wide. It was huge, and it was dangerous.

IN THE DOW TRUCK, Josh Wurman marveled at how prolific Storm A became, and then he became alarmed. Supercells flared one after the other around the region, and each began cycling rapidly through tornadoes, just as Fujita predicted. "It just seemed like a storm would go up and it made tornadoes and it made them over and over again." Sometimes supercells merge and create a vast squall line, expending their energy in the form of hail and rain. However, these supercells stayed as individual organisms. The DOW tried to keep pace with Storm A, and Josh watched those same twin twisters. The two DOW trucks had become separated, and each was now just trying to keep up with the tornadoes.

A6 turned into a dark wedge, its winds topping 150 to 200 miles per hour. The satellite, its seventh twister, made one orbit and disappeared. The main trunk continued churning toward the farming community of Chickasha, a town of 15,000 people more than forty miles southwest of Oklahoma City.

Gary warned viewers in Grady County: "If you haven't gone to the cellar, you really need to go now. This is a huge circulation. There are vortices everywhere. This is extremely dangerous, so you folks in the path of this tornado, get below ground. If you can't do that, get in the center part of your house, a closet or bathroom. Get on the east or north wall. Lots of pillows and blankets. Get in the bathtub. Put the kids in the bathtub, get on top of the kids. This is extremely dangerous. This is a major tornado of significant proportions. Wind speeds we don't know, but it's going to level most houses. You folks in the path of this storm, it is broad scale; it is broad scale, but do not expect a single funnel at all times."

Inside the Bear's Cage

• • •

Steve Chapman, Chickasha's emergency manager, watched the thunderstorm develop on the OK-FIRST monitor. Using OK-FIRST, the Oklahoma Mesonet, Chapman had dispatched volunteer spotters around the city an hour earlier. Chapman, like many other emergency managers, had a second job, director of the local airport, but he rushed to the No. 2 Fire House where the switch for the tornado alarms was located.

But he didn't need anyone to tell him a major tornado was headed for his city. He saw it on his computer screen. A spotter called in as soon as A6 came over the horizon.

"We gave it everything we had," said Chapman. A flip of the switch, and all eight of the city's sirens screamed. At the edge of the farm town, the tornado turned north.

"We dodged a bullet," said Chapman. Almost.

KWTV carried Ranger 9's aerial shots live for the entire nine miles that A6 was on the ground. One explosion after another lit up the tail of the tornado, as though it were bombing the countryside. The explosions were from electricity transformers being jerked from their moorings, either utility poles or homes. It injured nine people during its nine-mile jaunt.

"Grady County is reporting injuries with this tornado. This is a wedge tornado. This is the type of tornado that you need to be below ground level." Storm A heaved again, as number 6 gave way to number 8, almost its equal.

"We have a tornado on the ground again. Ranger 9 is over Chickasha so this apparently is a little to the north of Chickasha. My God! Talk to me, Val."

"Gary, it never even missed a beat. It's back on the ground again. It's just as big as it was before. It's a wedge again."

A8 clipped the city's small airport and ripped the roofs from two hangars, and whipped along for four miles. A mobile home park nestled at the edge of the airport took a hit. Minutes before the twister arrived, after hearing the storm sirens warning of A6, thirty mobile home residents scurried into an underground shelter. The tornado ripped off the cellar door but left them unharmed. A wing from a damaged plane floated upward, caught in the whirling winds of the storm's mesocyclone. It would be found later, lying amid rubble in Oklahoma City.

Gary again warned viewers: "It's outside the airport. Multiple power line flashes. Debris cloud in the air. Tornado moving through the airport in Chickasha. We have a very significant, possibly deadly tornado on the ground."

A8 disappeared almost as quickly as it came, covering only four miles. Storm A took a two-minute breather. Val and Amy scanned the black cloud, trying to find the next twister. The storm wasn't done yet, not by a long shot.

CHUCK DOSWELL HAD A late start on the chase. Known for his blunt talk and brusque manner, Doswell still was one of the nation's premier storm researchers even as he neared retirement. A project kept him holed up in his tiny NSSL cubicle all day. It wasn't until nearly 5:00 p.m. that he thought to stop by the Storm Prediction Center before leaving for home.

"I was somewhat shocked to see storms developing, and there was a level of concern that was considerably higher than it had been in the morning."

Doswell had been chasing thunderstorms for twenty-seven years. On his first glance at Storm A as it developed near Lawton, he wasn't sure there was much to get excited about. "It takes a lot to get me out the door."

But it was practically in his backyard. Roger and Rich passed him on their way out the door. "Why don't you guys go down there and kill this thing?" he shouted after them.

He drove by his house to pick up his camera equipment. He first headed north toward Oklahoma City but cut over to Interstate 44 and began driving southwest toward Chickasha.

A general rule of the storm chase is to stay off interstate highways. Tornadoes are hard enough to chase along spiderwebs of back roads. But the interstate highway limits chase options: there are fewer exits and limited escape routes. I-44 also was a turnpike with toll gates. As he headed toward Chickasha on I-44, he noticed people already stopping at the interstate's overpasses, some even climbing up the embankments for a better view of the storm coming into view. *What in the world are they thinking?* he wondered.

All over the metro area, people went about their business. The city's AAA baseball team, the RedHawks, was playing a doubleheader with Memphis at Bricktown Ballpark. During batting practice, the humidity was stifling, the sky cloudy but not ominous. More than 2,000 people filed into Bricktown in downtown for the opening pitch at 6:05 p.m. At Westmoore High School in southwest Oklahoma City, more than 400 parents and students showed up for an awards ceremony. In the suburb of Moore, dozens of residents were on hand for a critical city council vote on a bond issue.

Moore resident Charlie Cusack turned on his television, settling in to watch the live tornado coverage. Truckers at the American Freightways Company were returning to the terminal, finishing their shifts. Hundreds of people were doing some postwork shopping at Crossroads Mall despite the bad weather warnings. Katherine Walton and her son, Levi, were driving home to Anadarko on 1-44 after taking their neighbors to a car dealership in Oklahoma City. Had they not had car trouble, they would have been home already.

At tiny Bridge Creek, Chad Erwin worked on a genealogy project on his computer. At Ridgecrest Baptist Church, the last of sixty children in the day care were picked up by parents. Across the street from the church, Janie Pruett planted monkey grass in her backyard until she heard the state trooper racing up and down nearby 1-44, sirens wailing.

By the time Doswell got to Chickasha, it seemed the only cars on the interstate were storm chasers, probably fifty or sixty of them. He set up his tripod just as Storm A's eighth tornado fizzled. Within two minutes, another formed.

At 6:23 p.m., the tip of the tornado swirled the grass in a pasture near Amber, approximately thirty miles southwest of Oklahoma City. This tornado, the ninth from Storm A, started slowly, its winds hovering around 100 miles per hour. It quickly began to bulk up as the surface winds rushed toward it. It seemed to heave and grow.

Roger and Rich were behind the twister on a section-line road, firing off a few shots with their cameras. A9 turned into a wedge that Roger estimated was four times as wide as it was tall. Doswell was east of the tornado, still on I-44. Both scientists described an eerie "steady state" by the twister.

Unhurried, A9 marched at an even speed parallel to I-44. Doswell could not believe his good fortune. The monster twister gave him an uninterrupted view, and Doswell—along with about fifty storm chasers, two DOW trucks, and two vans full of Minnesota high school students—drove along beside it.

"I've chased an awful lot of tornados and often wondered what it would be like to see a real big nasty one, because, you know, I've seen my share, but never one of these outright giants."

Doswell had waited an entire career as a tornado researcher, nearly three decades, to see something this massive.

It took the Minnesota teenagers less than three hours.

Teacher Craig Wolter and his twelve high school seniors inched their vans behind the two DOW trucks as the first chase began near Lawton. For the first tornado, each DOW truck was steadied by hydraulic lifts so the radar could take clear pictures on either side of the tornado. Now they were just trying to take any radar picture they could.

"Oh, wow," the teens screamed in unison as the sixth tornado dropped down near Chickasha and dug into the dirt. The twister was on the ground long enough for Josh Wurman to stop, set up, and scan the storm. The students piled out of the vans and again peeked into the extended DOW cab as Josh's computer screen displayed the reds, yellows, and greens of hard rain and even harder winds.

Josh took a gamble and the DOW trucks drove onto I-44 heading north, just as A9 was forming. Like Doswell, Josh was in the best place on the worst road—an interstate turnpike.

Gary England had now been on live for nearly two hours. He watched a video monitor carrying Ranger 9's aerial shots.

"We see the whole darn thing coming down again, Val."

"Yeah," said Val. "And it's headed in a bad direction." A9 was cruising toward the northeast, toward Oklahoma City.

FORECASTERS AT THE NATIONAL Weather Service in Norman settled into an eerily silent chore. Unlike Val and Gary, there was no demonstrable excitement to their voices as they issued one tornado warning after another. NWS began calling in off-duty staff. The outbreak was becoming much bigger than they had imagined.

The Norman office usually had one person on the warning desk. By the end of the evening, there were four people. So massive was the outbreak of supercells and so prolific were the storms that they were issuing tornado warnings every four minutes. Such a rapid flow of information would have been impossible just a couple of years earlier. At least the weather office in Norman had all the modern technological tools. The Doppler radar provided the pictures. The AWIPS computer allowed the forecasters to divide the state into sections so that each meteorologist could keep an eye out for only his quadrant. The SKYWARN spotters called in with each sighting.

Never before had one weather office been so inundated. The massive supercell outbreak gyrated across central Oklahoma. Because of the new equipment, the tools that politicians hadn't wanted to buy, forecasters were able to give the public an average warning lead time of twenty-one minutes.

As A9 moved within ten miles of Norman, the National Weather Service and the Storm Prediction Center prepared to abandon ship, activating their emergency evacuation plans. If the tornado turned right, as many do, it could swirl into Norman. The NWS forecasters called the

Tulsa weather office to brief them on the status of the central Oklahoma supercells so Tulsa could take charge. The Norman forecasters would stay at their posts as long as they could.

At the Storm Prediction Center, off-duty NSSL researchers and SPC forecasters filtered into the concrete building, some out of curiosity, some seeking safety in the bunkerlike construction. The Storm Prediction Center made plans to turn over the nation's weather safety to Offutt Air Force Base in Nebraska, headquarters for the nation's Strategic Air Command. Offutt years earlier had assumed the operations of the Air Force Weather Agency, the old operations once headed by Fawbush and Miller at Tinker Air Force Base. The Weather Agency remained the military's equivalent of the Storm Prediction Center.

The Storm Prediction Center extended weather watches from Oklahoma to southern Kansas. There was nothing more for them to do but stare at their radar screens in amazement.

"It's COMING RIGHT UP alongside the turnpike. The circulation is about six miles across," Gary told viewers. Explosions rippled at the tail from downed power lines. "Oh man, oh man, oh man. Jiminy Christmas. You folks watch this."

Tall, lean, middle-aged Bridge Creek school superintendent Terry Brown took his wife and son out to dinner in southwestern Oklahoma City, just across the Canadian River. It was his son's birthday. Brown kept jumping up, running to the bar to see the twister being broadcast live on television.

"You've seen tornadoes before. We're trying to have family time," his wife chastised him.

"I've seen tornadoes before but never like this. I think it's coming straight at us. We might want to think about getting out of here," Brown told her.

Val cut into the KWTV broadcast. "We're starting to see residential sections in the path! It's starting to come into the outskirts of the metro area!"

"This is a very extremely serious storm. The velocities are very high," Gary said. "It is a wedged-shaped tornado. It's moving into the metro area, densely populated areas."

DOSWELL MARVELED AT THE steadiness of the tornado. It showed no signs of fizzling. Most tornadoes are on the ground for fifteen minutes or less before they unravel. A9 only grew bigger.

"This storm was just bigger and nastier and more powerful than any other I had ever seen. And then you had this large, wide tornado on the ground for an extended period. My meteorological experience suggested to me that this one wasn't going to dissipate real easily or real soon."

From Ranger 9's aerial shot, Gary could see the 1-44 tollbooth. The tornado was still moving parallel to the highway. "The mesocyclone is cranking up, my friend. It is becoming stronger and stronger and stronger. It's going to move into Moore and parts of south Oklahoma City. This is a deadly tornado."

Val and Amy were being pounded by trash spewing from the twister.

"We can hear the tornado right now. It is roaring. There's debris. We're just about to get hit with debris. We're going to have to retreat, Gary. A mile outside of the circulation it is dropping debris."

Gary tried again to rattle viewers into action.

"Let me tell you, folks, this is deadly, deadly serious. Take your tornado precautions and do it now. It is moving toward west Moore and southwest Oklahoma City. It has left an incredible trail of damage. It continues to be an intense, possibly deadly tornado. You need to be in below-ground shelter. As you look at it, this tornado is moving through parts of Newcastle and southwest Oklahoma City."

At the restaurant, Terry Brown saw the turnpike gate on the television report and feared the tornado was headed straight for the Bridge Creek school. "We need to get in the car and we need to leave," Brown told his family. They headed west so they could get away from the northeastward-moving twister and then doubled back around so Brown could get to the school district office.

"We weren't on any maps," Brown said. "The TV stations, and they did a wonderful job putting out warnings, but it was 'south Tuttle' or 'north Blanchard.' To those of us in Bridge Creek, south Tuttle is west of us and north Blanchard is east of us. I thought there were going to be hundreds of deaths out here. I was just afraid people wouldn't understand it was coming straight for them."

Both DOW trucks trailed A9 by about a mile. "Our goal was to get ahead of the storms and set up this triangulation again as they came toward us. But we could never get around it because there were lethal winds on I-44," Josh recalled. The DOW trucks repeated their stop-and-start scans as they moved north on I-44. Once they got within a mile, they stopped, did a Doppler scan, and tried to catch up with it again.

At the Weather Service, science officer David Andra, an unflappable Kansan with a clinical demeanor, called his wife: "Put the cats in the crate and get them in the car. Be prepared to drive south as fast as you can." Getting his wife to take action was easy. He wanted others in the path to do the same.

He took one final stab at conveying the danger: "Tornado emergency," he wrote at 6:57 p.m. in a public warning for the media, emergency managers, and NOAA Weather Radio. The term isn't in the NWS handbook. He made it up on the fly, trying to reinforce in the media the dire situation. The agency that had been reluctant for decades even to mention the word *tornado* out of concern for public panic was now trying to create as much fear as possible—so that people would take some steps to protect themselves.

"I wanted to grab attention and characterize the nature of what was about to happen: that it was going to be a catastrophic event, it was basically an emergency situation in the metropolitan area; and that as an emergency, people ought to take the right action, take it soon or they may not survive. And there aren't many cases in your day-to-day life where someone comes on the radio and says, "What you do in the next ten to twenty minutes may make a difference between whether you or your family survives. Pay attention."

Tornado emergency in south Oklahoma City metro area. . . . At 6:57 PM CDT . . . a large tornado was moving along Interstate 44 West of Newcastle. On its present path . . . this large damaging tornado will enter southwest sections of the Oklahoma City metro area between 7:15 and 7:30 PM. Persons in Moore and South Oklahoma City should take immediate tornado precautions! This is an extremely dangerous and life threatening situation. If you are in the path of this large and destructive tornado . . . take cover immediately. Doppler radar indicated this storm may contain destructive hail to the size of baseballs . . . or larger.

"We wanted people to understand this was a life-threatening emergency, and you couldn't sit and watch television, you couldn't go shopping, you couldn't spend time talking on the telephone—it's a high threat and it was highly likely to be fatal," Andra said.

At the Storm Prediction Center, Jeff Evans stared at a red blob on the Doppler radar screen. Red reflected the intensity. The mesocyclone maxed out, stretching across his monitor. His stomach churned. He felt sick just watching it.

"I thought I was looking at the first tornado to kill a hundred people or more in nearly fifty years. I knew that area; it was nothing but housing additions."

There was no television in the center. But now Jeff was getting secondhand reports from center meteorologists and lab researchers. A few of them climbed to the rooftop, seeking a once-in-a-lifetime view of nature at its most dangerous. All they could see was a black wall of clouds. That's all anybody could see now. The twister was enveloped in wind and rain, the bear's cage.

THE HAIL CAME FIRST, the size of baseballs. Chad Erwin snatched a piece of carpet from his porch and ran to his car to cover the front windshield. "You're nuts!" his son yelled after him. Hail pounded his head until he made it back indoors. His teenager watched the storm through a back window. "I don't like the way that looks."

STORM WARNING

Chad peeked out the back door of his double-wide. To him, it looked like a black curtain along the horizon. As the tornado approached Bridge Creek, it was more than one mile wide. It was so big his son didn't recognize it as a tornado. "He said, 'What is that?' And I said—I just stood there for a minute and a little man in the back of my head, he goes, 'It's time to get your butt outta here.'"

The black curtain appeared to lift momentarily, appearing to jump a blue cylindrical water tower about a half-mile behind his house. He yelled at his wife to put on boots and screamed at his children to grab the two dogs.

There was a double metal culvert in a steep ravine 300 feet from his house. It formed a low bridge for a gravel road above it and allowed water from the ravine to flow into a holding pond. Chad and his family lived on one side of the ravine; Kara and Jordan Wiese lived on the other.

Chad, his wife, and his two kids made a run for the steel culverts—the wind so strong at this point that they could not run in a straight line. The children and dogs went in first, crawling through the three-foot opening. His wife followed. Chad watched the tornado until it reached the fence line at the back of his five acres. He dropped to his knees and scurried in last.

Kara and her mother were talking again on the phone. She asked if Kara wanted to come to her house.

"She wanted to leave but she was scared to leave because there were so many tornadoes. They would talk about the tornadoes headed for Newcastle and the tornadoes headed for Tuttle. She was in the middle. She was scared of getting in the car. They had a very deep ravine behind her house that she said they were going to get into."

The phone went dead. Nothing. Not even a dial tone. And Mary's heart began to pound with dread.

Hail pounded Kara's trailer like cannon fire; it flattened the flowers and dented her car. She saw the Erwins running toward the ravine. Kara wrapped Jordan in a heavy winter coat, the best armor she could find in the closet. The bullying winds that were racing into the tornado rocked the mobile home. The newly planted trees in her front yard bent back and forth. The sky was black now, dark from the rain and the hail and the massive mesocyclone that hovered overhead.

Jordan held tight to his mother, his skinny arms and legs wrapped around her body with all his might. They were going to make a run for the steep ravine just as the Erwins had. But she changed her mind. Perhaps it was the hail; perhaps it was that unmistakable sound, a roaring shock wave of deafening wind.

They sat in the bathtub facing each other, or they were running down the hallway to the bathroom to get into the tub; recollections faltered. A monstrous wind toyed with their mobile home, teetering it back and forth. Kara squeezed Jordan harder and harder; she would never let go. Not ever. The ties that anchored the mobile home snapped, and it rolled off from its undercarriage. Everything was airborne at this point, tumbling end over end. The furniture, the collection of unicorns, the toys in the back bedroom, her college application papers. The great wind lifted and tore and swirled. Jordan held Kara as tightly as he could. He felt her slip from his grasp, and he was alone, tumbling into a blackness that enveloped him, and it was as if he went to sleep.

DEBRIS STARTED TO RATTLE around the opening, and Chad put his feet up to try to block it. A sheet of green carpet—which just seconds before was in his neighbors' living room—snagged a tree limb outside the culvert and flapped against the opening. "I thought it was going to come in and clean us out like a pipe cleaner," Chad said. Instead, it acted as a door, shielding the Erwin family from the missiles.

Looking out the other end of the sixty-foot culvert, all Chad could see was red—the red clay being scoured up by the winds. Neither could he hear anything. It was almost like a vacuum was created inside the culvert. "Everybody keeps talking about a freight train, but we couldn't hear nothing." But they could feel it. They could feel the ground vibrating.

STORM WARNING

When the vibrating stopped, Chad took his pocket knife and cut a slit through the carpet so he could get out. In the distance, Chad could see the black curtain still moving through Bridge Creek; it appeared to stop and take a slight right turn, directly into the Ridge Crest Baptist Church. Two day care workers huddled in a church closet as the tornado tossed the new church van into the pulpit. The winds tore the church apart. Only the closet remained.

As it crossed through Bridge Creek, the tornado grew again to more than one mile wide, appearing as though the entire mesocyclone dragged across the ground. It was so massive that several people, like the Erwins, did not recognize it as a twister.

Tom Tinneman heard Gary England's warnings that people should get below ground, but he had no shelter at his mobile home. He and his two girls ran into a closet, and he felt the home tip over and begin to roll.

Nearby, nineteen-year-old Amy Crago held her ten-month-old daughter, Aleah, as she, her husband, and her parents sought shelter in the closet of her parents' wood house. The vortex exploded the home, catapulting Amy and Aleah into the air and in opposite directions. An outside wall fell on her mother. Her father held her mother's head as a single tear rolled down her cheek.

Down the road, Deon and Samantha Darnell, new parents of a three-week-old boy, took shelter at his parents' home. The winds swept the house off its foundation and carried the entire Darnell family away. Baby Ashton and his grandmother were missing amid the debris.

The Bridge Creek ravines provided the illusion of safety for several people. Wesley Early, who the previous year dumped thirty tons of gravel to level his five acres, ducked into a ditch as rocks pelted him. But the winds were too strong; they picked him up and hurled him 100 yards.

DOSWELL STAYED WITH A9, following slowly behind mile after mile. Either side of I-44 was mostly pasture that slowly gave way to large commercial sites for stock trailers and oil supplies that in turn gave way to small shopping outlets and, once across the river, housing additions. Doswell, in his log, described a "giant mulching mower," spewing debris everywhere.

At first, it failed to register that the twister was headed straight for Oklahoma City. Doswell thought it was still in open country, though he could see the power flashes from exploding electrical transformers.

The rain turned pink. Wisps of pink housing insulation began to fall on Doswell's car. "Someone is getting hammered," he thought. It was Bridge Creek.

SCOTT PITTMAN AND TWO friends were driving south of Interstate 44 when he saw the tornado moving beside the highway. He did a U-turn, heading back north on the shoulder of the southbound lanes. Jersey barriers prevented him from getting in the northbound lanes, so he stopped at the first overpass. He and his friends scrambled up the embankment.

They were soon joined by Kathleen Walton and her eleven-year-old son, Levi. The Waltons had been driving to their home in Anadarko, an hour south of Oklahoma City. If not for the balky car, they might have had just an interesting tale about seeing a tornado. Instead, they were now hiding from it. Kathleen and Levi scurried up the concrete slope as the first wave of wind attacked. Pittman grabbed Levi's hand and pulled him up to the girder.

A9 moved directly across the I-44 underpass.

Levi, not Kathleen, had a grasp on the steel girders; Kathleen held her son's hand. A second blast of wind began pulling her down the embankment, dragging her son with her. The wind tugged on her as Levi tried to pull her back. She told him to close his eyes. "I love you," she shouted above the wind. And she let go.

The overpass became a wind tunnel filled with glass, mud, lumber, gravel, and trees. Pittman's leg was sliced to the bone by a highway sign. And then the wind stopped. The boy began to cry, "Where's my momma?"

STORM WARNING

• • •

Josh's DOW truck stopped on I-44, and they did a quick scan as the tornado roared out of Bridge Creek and headed across the river.

The wind speed was 318 miles per hour a few feet up. He estimated the surface speed at 301 miles per hour. No wind that powerful had ever been recorded at the Earth's surface.

The force of wind followed a simple mathematic formula: force equals velocity squared. A 300-mile-per-hour wind was not three times as powerful as a 100-mile-per-hour wind; it was nine times more powerful. The most powerful tornado ever recorded was taking aim at the residential areas of southwestern Oklahoma City.

Craig Wolter and the Minnesota students, still trailing the second DOW truck, slowed because of the winds and rain. The vans shimmied along the highway, and Wolter feared they could no longer keep up with the heavier DOW trucks. He broke off the chase before the twister entered the metro area.

"It's approaching the river," Gary told viewers. "The sirens are going off in Moore. It's moving northeast to the Moore area. This is a long-tracked tornado, potentially deadly. The wind speeds are quite strong, we fear. You have time, you still have a few minutes in Moore to move to a place of safety, but not much."

A9 was moving at full speed into densely packed neighborhoods.

The tornado crossed I-44 and the South Canadian River. Dead ahead was Will Rogers World Airport, the region's main civilian airport, where hundreds of people were stranded as the airport waved away aircraft. Nothing could take off or land in this weather. The tornado took a right turn and entered a densely packed neighborhood just before officials evacuated the terminals.

The twister, its tail contracting and expanding, plowed into Country Place Estates, an addition of modest single-family homes and apartments. Eastlake Estates was next. The housing addition of smaller patio homes was only seven years old and home to many retirees. The tornado was smaller now and slightly less powerful, but still fierce. It was a half-mile wide when it tore through the housing addition, leaving stubble and empty concrete slabs that had once been foundations for homes.

At Westmoore High School in Oklahoma City, more than 400 parents and children gathered for an honors ceremony. A quick-thinking assistant principal ordered the people out of the auditorium and into the bathrooms and reinforced hallways. They hunkered down next to the school lockers and turned their bodies away from the windows.

The twister's edge caught part of the school. It hurled cars from the Westmoore parking lot into the neighborhood homes across the street.

Josh and his crew were ninety seconds behind the twister, trailing it through residential side roads. It was an eerie scene of absolute silence. There were no people, no sirens, just the distant roar of the tornado. "Didn't there used to be houses here?" someone asked.

Now A9 was moving due east. Julie Rakestraw crawled into the closet of her Moore home and covered herself and her child with blankets. Her home exploded around her. A chunk of a two-by-four entered near her right ear, through her neck, and into her left collarbone. She knew she was in trouble when her neighbor, a paramedic, tossed aside the debris to get to her and her child. He took one look at her and the twenty-inch piece of lumber sticking out from her neck and screamed, "Oh, shit!"

Charlie Cusack sat on his couch and watched the tornado on television. His wife, Pam, was preparing tacos to take to the monthly Girl Scout meeting. "Let's turn on the TV and see the weather. You may not want to go," Charlie told her. The previous October, the day of the autumn outbreak, a tornado had come perilously close to the Cusack home. Its winds damaged his fence, knocked down tree limbs, and over-turned his daughters' trampoline. Charlie assumed May 3 would be a repeat of October—close but nothing to get excited about.

The big twister already was being broadcast on all the local television stations. Helicopters from Channel 9 and Channel 4 captured dramatic footage. For more than an hour, he watched it chug parallel to I-44, watched it tear through Bridge Creek, watched it cross the river, and watched it edge eastward, toward his home.

"The stupidest, stupidest thing that I've ever done in my life was turn on the TV, sit down, and watch that thing from about 6:15 p.m. until it came in the front door," Charlie said.

Charlie surfed the various local channels, flipping back and forth between Channel 9 and Channel 4. A little after 7:00 p.m., the storm sirens sounded throughout Moore, and an emergency broadcast interrupted all television stations to urge people to seek shelter. A glitch in the system prevented the television broadcasts from resuming.

Charlie said it sounded like a busy telephone signal, like someone forgot to hang up the phone. He could see Gary England's lips moving, but there was no audio. He figured this was about the time England was telling people to get underground or die.

But it was too late for the Cusacks to leave. Charlie looked out the front door to see hail falling outside and a menacing black cloud. Charlie gathered his two daughters and wife, and they headed for a first-floor bathroom.

He heard the electrical transformer at the end of his block explode. For an instant, the house went dark and quiet. "And then there was that noise," the sound like a space shuttle taking off.

Charlie leaned against the hollow bathroom door as it bucked from the wind. His wife and daughters were in the bathtub. He had put on a leather jacket to protect himself and now thought to himself how silly that was. The leather jacket would provide no protection from this twister. The whole house began to shake. "It was like a can of peas in a grocery cart with a wobbly wheel. The house was shaking like that. My wife said she thought the whole thing was going to take off and go any second."

The sound became deafening, the roar of the wind combining with the dismantling of the Cusacks' home. There was a sickening screeching noise, like the metal-on-metal sound of cars colliding over and over and over again. And there was a sweet scent that reminded Charlie of freshly cut Christmas trees. The aroma of the stripped vegetation seeped in under the bathroom door.

"It's time we start praying," Charlie said.

"What do we say?" asked his wife.

"Hail Mary full of grace."

And then it stopped.

Charlie shoved open the bathroom door. A ceiling fan leaned against it, making it hard to push. It wasn't Charlie's ceiling fan; it was a neighbor's. Charlie's ceiling fan was gone, along with most of the ceiling. The tornado had sheared away the second floor of the Cusacks' house and caved in the living room walls. Only the back portion of the home, where they had taken shelter in a bathroom, was left standing.

As he walked outside the bathroom, it took a moment for the scene to register. There was nothing there. They picked their way through the debris to get out of their house. They walked out an opening where the living room window had been.

"Oh, my God," Charlie said as he looked across the street at Kelly Elementary School, which his youngest daughter attended. There was nothing there taller than a couple of feet. The school was gone. Splatters of sticky tar were everywhere. The roof of the school had been retarred the previous week.

IN DOWNTOWN OKLAHOMA CITY, the RedHawks halted play during the second inning. An announcer calmly asked the patrons to leave their seats and move to the stadium basement. Jack Damrill, doing play-by-play for the home team, remembered a chaotic evacuation as stadium officials sought to clear the stadium quickly. They opened the field, and stadium personnel began directing the fans through the dugouts, through the lower levels, through any opening that would get 2,000 people underground quickly.

He could see the sports reporters watching a television in the next room but could not hear the audio. "The guys were just going nuts on TV so I knew something was up," Damrill said. He and a few other reporters scampered to a deck behind the press box. "All you could see to the south was just pitch black."

A9 was churning through Moore, nearly ten miles south. Pieces of pink housing insulation floated like feathers onto center field. The fear was that the tornado would move into downtown Oklahoma City itself. Bricktown's underground storage area was the safest place in the park. "That's where they tried to get everyone, because obviously it was underground and you kept hearing the weather guys on TV saying, you know, 'You do not need to be above ground if you're in the path of this storm.'

"The storm was so massive all you could see was this huge black cloud that went all the way from the sky to the ground. And you knew a tornado was in there, because you could see it on TV and you could hear everyone talking about it on TV. But all you could see from our vantage point was just the one huge monstrous black cloud," Damrill said.

From the upper deck, they could see the black cloud moving across Interstate 35, the second major highway it had crossed. That's when Damrill headed for the basement, running to the RedHawks locker room, where he watched the tornado on television.

"THIS IS A HUGE tornado, deadly tornado. A very wide-path tornado. It is so large we can barely see it," Gary told viewers. "Civil Defense is calling in all staff in the metro area."

Traffic on Interstate 35, the nation's main north-south artery in the heartland, had ground to a halt as people scrambled out of their cars and took shelter under the Shields Boulevard overpass where Brian Hanson was trapped in his pickup. He could see the twister coming directly toward him but had nowhere to go. People left their cars in the middle of the highway, blocking all traffic. Brian, the deputy emergency manager for the city of Moore, was trying to get to city hall. He had tried to wedge his pickup past the parked cars but now could go no farther. And neither could he open the pickup doors to get out.

The slimly built Hanson crawled to the pickup floor and hunkered down. The winds rocked his truck. The windows exploded, showering him with glass shards.

Tram Thu Bui and her husband, Thanh Pham, were making a surprise visit to relatives in Oklahoma City. Only three months earlier, the young couple had opened their own restaurant, Golden Palace, in a nearby town. They had their three-year-old daughter and two-year-old son with them. Pham stopped at the Shields Boulevard overpass and carried his two children to the girders just as the high winds hit. He turned to look for his wife. She was gone.

THE MOORE CITY COUNCIL had gathered at 6:30 p.m. at city hall for its weekly Monday meeting. Gayland Kitch, the emergency manager, didn't stay for much of the meeting. He secluded himself in the tiny niche of an office on the first floor that served as the city's emergency headquarters.

A three-story brick building, the Moore City Hall had an airy atrium with three levels of windows that faced the east and large windows at either end of the building. Relatively new, it opened at 301 West Broadway in 1986. A large state-of-the-art siren was just outside the north doors.

After Kitch sounded the alarms a little after 7:00 p.m., city council members and dozens of citizens headed for the restrooms on the first floor. They emerged when it was clear the tornado would bypass city hall and watched the huge, black cloud drag across the ground.

The tornado skirted the edge of the huge First Baptist Church of Moore and moved into open land. The tornado's debris field was so large that even 300 yards away, it thoroughly scoured the back side of the Moore water tower, removing all the paint.

A9 now carried so much lumber, trees, cars, and whatever else that the debris field itself became visible on the Doppler radar as tiny dots.

This was the tornado, Gary thought. This was the one he knew would someday visit the metro area, and all he could do was keep repeating himself: "Get below ground." The closets and bathrooms, usually safe shelters, could not withstand the wind power. People had to be out of its path or underground.

"You folks in the Tinker area, south Del City, take immediate tornado precautions. Huge tornado in the Moore area. Damage is severe. Get below ground. You folks sitting through this storm get below ground immediately. This is absolutely incredible. I tell you, this is deadly serious. We've been talking about it for an hour or two now. The damage is massive."

At the National Weather Service, David Andra again issued a blunt warning: "Persons in southeast Oklahoma City and Midwest City are in danger!"

AT THE INTERSECTION OF two major interstate highways, I-35 and I-240, hundreds of people were shopping at Crossroads Mall's 125 stores. Mall managers evacuated people to storage areas in the basement and kept them there for ninety minutes, until they were sure it was safe. The twister passed a few blocks east of the mall, aiming at the American Freightways terminals.

Trucker Anthony Batagglia saw it coming from a distance. The San Francisco native called his pregnant wife to tell her he was seeing his first tornado. Batagglia was working the city routes on May 3 so he could be at home every night with his family. Son Antonio Mac was only two. They were expecting another baby, a girl, within weeks.

Batagglia joined several other drivers and secretaries as they huddled in the northwest corner of a loading dock. Batagglia and trucker Jerry Webb, both big men, spread their arms and used their bodies to shield the others.

The twister hit the terminal square head-on. It shredded the heavy semitrailers and tractors as though they were made of paper. It pushed one trailer up a ramp and onto the dock, crushing the two drivers and injuring several others.

The tornado turned left, taking a more northward track, once again crossing back into housing additions.

Gary kept repeating himself, changing only the names of neighborhoods as A9 churned across the metro area and then took another turn, to the north. "Large, large long-tracked tornado continues. It is moving to Del City, Tinker, Draper Lake. Oh, boy, I don't believe this."

CAROLYN STAGER, A LOBBYIST for the Oklahoma Municipal League, originally intended to have dinner with a friend Monday night. Her mother called to tell her that it was supposed to storm. She cancelled her plans and headed for her home in Del City, located on the outskirts of Tinker Air Force Base.

Carolyn called her daughter, who was seven months pregnant and whose husband was working late. Carolyn urged her to come to her Del City home, a two-story brick house. "We'll be safer," Carolyn told her daughter.

Just after 7:30 p.m., her ex-husband, who lived in Moore, called to say a tornado had just passed through the area. "You need to take shelter," he told them.

Carolyn looked outside as the black mass covered the entire sky. She realized it was too late to get in the car and drive away to safety. She and her daughter crawled into a closet under a stairwell.

She remained conscious the entire time. She heard the windows explode. The house began to fly apart. She felt herself being sucked out of the closet. The tornado ripped the carpet and the tile off the floors, out from under them. Carolyn remembers being tossed and spun from one side to another, her body being battered by flying debris.

"There was a time I thought I was dying. I knew my body couldn't take much more of the abuse."

And then it was over. Carolyn lay limp on the ground, alive but barely awake, her pelvis broken.

Her scalp had been sliced from her head.

Her daughter looked as if she had been mauled by an animal; divots of flesh had been torn from her limbs. But, she, like her mother, was at least alive.

The tornado clipped the edge of Tinker Air Force Base and moved into Midwest City. A9 scattered thirty cars awaiting service at Morris' Auto Machine and Supply, hurling one empty school bus more than 200 yards. Weary travelers already had checked into a series of motels— a Comfort Inn, a Clarion Inn, and a Hampton Inn—along I-40. Guests hid in the hotels' bathrooms as the twister rumbled overhead. Several Hampton Inn guests made a dash for the Comfort Inn lobby next door. A9 ripped the second story off the Hampton Inn, piling cars atop the new one-story building.

Mike Pederson, manager of the Cracker Barrel restaurant along the interstate, saw nothing but flying debris when he looked out the window. He herded thirty customers into the restaurant's kitchen, which was about all that remained standing after the twister. A Penske rental truck crashed through the roof of the dining room.

A9 headed directly toward another shopping mall north of I-40 when it just stopped. The twister feathered into a ropy funnel, its vortices untwining like the pulled threads of a string. One hour and twenty-five minutes. That's how long A9 stayed on the ground. Its thirty-eight-mile path was easy to find: there was nothing but destruction.

Storm A still wasn't finished. It produced five more tornadoes, including one that tracked seven miles and injured two more people.

A9 ALONE WOULD HAVE been a historic event. But the atmosphere was in full frenzy. The wrestling air currents that spawned Storm A exploded into ten more supercells, and each mesocyclone spun out a tornado of increasing violence.

A satellite picture showed a massive cloud cover over the state, with numerous little volcanolike puffs of cloud tops. The supercells extended into the stratosphere.

For many of the emergency managers, trained to use OK-FIRST, the night was just beginning. This was an event for which Ken Crawford had envisioned OK-FIRST. Once the giant twister rocked Bridge Creek, Oklahoma City, and Moore, the local television stations turned their cameras on the devastation. Tornadoes were still forming rapidly, and Gary England continued his on-air warnings. But the news in Oklahoma City was now the deaths and destruction.

Storm D formed near the tiny town of Pink, southeast of Oklahoma City. A small tornado spun around for a few miles and died, but it quickly reloaded. It was nearly 9:00 p.m., but the supercells and their twisters still pounded the state.

Seminole County emergency manager Herb Garner was manning the OK-FIRST computer. The most forceful portion of Storm D was some twenty miles away, in the neighboring county. But Herb also knew a caravan of county police cars and ambulances was on I-40, heading toward Oklahoma City to offer aid and assistance to A9's victims.

He radioed them to halt in their tracks. The tornado passed in front of them, pushing two passenger cars off the road, killing Storm D's first victim. "That was the best call I've seen in thirty years of forecasting," Crawford said. The twister could have killed even more people.

Storm D continued moving northeastward, spewing one twister after another.

"All the TV stations and radio stations started reporting the tornado as a weather event coming toward the city. Once it hits the city, it becomes a news event, and all of our warnings stop. For those of us who are still downstream, this tornado is still moving along," said Ben Springfield, emergency manager for Lincoln County, midway between Oklahoma City and Tulsa.

Springfield, who enlisted numerous spotters and volunteers during severe weather, stationed himself in southern Lincoln County and quickly spotted a tornado on the ground from Storm D. He radioed a volunteer manning the OK-FIRST computer. Together, they plotted the likely path of the twister. It was taking dead aim at Stroud's tiny, twenty-five-bed hospital.

Springfield called his deputy, Joey Wakefield, who worked full time as a paramedic and was stationed at the hospital. "That tornado is heading straight for your hospital," Springfield warned him.

Wakefield, a strapping twenty-six-year-old, and his EMT partner helped the nurses move eight patients still on their gurneys into the surgery hallway, where there were no windows. A two-hour-old baby was placed in his mother's arms. They covered everyone with all the blankets they could find. Wakefield had been chasing tornadoes since he was eighteen. With the patients secure, he and his partner ventured outside.

"I turned around to tell him I thought we were going to be okay if the wind didn't shift on us. I no sooner said that than the wind shifted. A tree went flying by." They hustled back into the surgery ward. "They made fun of me afterward, but I went running into the surgery ward and shouted, 'It's here!' and dove under a gurney." An older nurse was already under the gurney. Wakefield wrapped his long arms around her to protect her, though later he said she was protecting him because he was scared to death.

"It felt like your whole head was going to pop. I thought we were going to die. There was a loud banging and all the ceiling tiles went up and came back down. You could hear glass breaking everywhere. I thought the walls were going to explode. The power went out. We heard this god-awful noise going across the roof." The tornado had lifted an outbuilding and carried it over the top of the two-story hospital.

The rumble lasted maybe fifteen to twenty seconds. And then it was silent. The ambulances were damaged, all the windows were broken, and an oxygen bottle was impaled in the side door of one ambulance. Storm D's twister moved a little farther north and hit the Tanger Outlet Mall head-on, scraping away the entire complex of stores. The only thing left was the concrete foundation.

The mall had closed an hour earlier.

WURMAN AND HIS CREW had driven almost to Tulsa chasing one supercell tornado after another. As darkness fell, they turned around and headed back toward Oklahoma City. Roger and Rich had left Storm A when its big twister headed into Bridge Creek and the metro area. They had found a tornadic gold mine in Storm B that arced to the west of Oklahoma City.

Inside the Bear's Cage

The VORTEX-99's mobile Mesonet crews had been chasing Storm B early on. "There were so many tornadoes we couldn't count them all," recalled a VORTEX volunteer. Because of the danger of chasing at night, the VORTEX crew was ordered home as dusk fell.

Roger and Rich had caught up with Storm B as it produced its third tornado and followed it as it moved west of Oklahoma County and made a wide, arcing loop around the capital city.

Like Storm A, Storm B grew stronger with each tornado. Roger and Rich were south of it. Wurman saw it on the Doppler and moved to the east of it. At 9:25 p.m. Storm B put down its twentieth tornado in a field near Cimarron City.

It was massive from the start. Gary had only one storm tracker filming Storm B. Ranger 9 had landed after A9 dissipated. Val had been caught up in the destruction in the metro area and quit the chase. Thousands of lives already had been turned upside down. It was dark. Only the lightning flashes illuminated this twister. But this tornado was a monster in its own right.

Gary saw the twister on a monitor. "You couldn't get the whole thing in the viewfinder." Wurman couldn't believe what he was seeing on his computer screen: the interior of the tornado was more than a mile wide.

Roger and Rich pulled off on a dirt road and watched the twister form. "Rog, that is one damn big tornado." It was six miles ahead of them, but they could see a massive base and the funnel lowering to the ground.

"The thing was twice the size of the tornado we saw at Bridge Creek," said Roger. "There were four bright power flashes, and they didn't illuminate either edge of the tornado because it was too big. We knew we were dealing with a beast. It was the biggest tornado either of us had ever seen, and we've seen a few real bruisers."

The monster was two miles wide. The mesocyclone circulation base, the hook echo, was ten miles across, 40 percent bigger than Storm A.

B20 stayed on the ground for thirty-nine miles, one more than

A9. Mulhall, a small town in Logan County, just north of Oklahoma County, was not much more than a bank, a store, a water tower, several churches, and lots of older homes. Logan County emergency manager John Lewis watched B20 on his OK-FIRST screen and notified deputies. The town had no tornado alarms. Two officers in police cruisers tore up one street and down another, blasting their sirens and warning residents of the approaching tornado. They continued warning people even as B20 bore down on them. They stopped only when one car was hit by a power line and the other by a tree. Neither man was injured.

B20 destroyed most of Mulhall by barely sideswiping the town. Jerry Patterson, vice president of the bank, took shelter in the bank itself. "The walls shook like Jell-O," he said. The twister picked up the local church and set it down atop a nearby house; it toppled a 400,000-gallon water tank, flooding nearby homes.

Other than Mulhall, B20 stayed in open farmland. It killed one man who had parked along the interstate highway, near an underpass.

"By far, the potentially most damaging and most spectacular, in a bad way, was the Mulhall tornado," Josh said. "The Mulhall tornado was the largest we had ever seen."

Within B20 churned several suction vortices, which themselves were at F4 levels. It was a tornado wrapped in other tornadoes, just as Fujita imagined. Scientists measure tornadoes by their internal core flow, which is smaller than what is visible.

B20 temporarily put Josh Wurman in the Guinness Book of World Records for the widest tornado ever recorded: 5,250 feet.

BY THE TIME B20 set down, Dustin was walking through a muddy field searching for Kara and Jordan. Cars littered the area, crushed like flimsy aluminum cans. He peeked into each one, trying to see if Kara was there.

Bridge Creek had become chaotic. People called out the names of neighbors. A helicopter hovered overhead. Police and ambulance sirens raced up and down I-44. But when it was momentarily silent, Dustin could hear a hissing sound like angry snakes coming from every direction. He paused and listened, and then it dawned on him. It was the sound of propane escaping from broken gas lines, lines connected to homes that were no longer there.

9

MR. TORNADO SEES HIS FIRST

TED FUJITA'S QUARRY ELUDED HIM. By the 1980s, Fujita had established himself as the premier tornado researcher in the world. Using only his eyes and his imagination, Fujita almost single-handedly charted the course of storm science, forcing researchers to rethink previous conclusions. He made discoveries—whether it was the "family" of tornadoes, the names and meanings of the storm clouds, multiple vortices, or the Fujita Scale—that made much of the meteorological establishment at first grumble at his observational methods, but later accept his theories as fact. And yet, Mr. Tornado had never seen a tornado.

It was no secret. The vanity license plate on his car broadcast the irony: TTF0000. TTF stood for Tetsuya Theodore Fujita, the zeroes for the number of twisters witnessed. Perhaps it was a signal of his surrendering the prospect of ever seeing a twister in person. Not that his interest in tornadoes had dimmed, but his research funding was pushing him in other directions. The nuclear power industry and the potentially destructive powers of a twister on a nuclear plant had been driving forces behind tornado research, but the emphasis on nuclear power as an energy source had waned in the 1980s.

"A success in research signals the onset of funding recession," Fujita quipped. "Unless one moves into a new research area, the recession could turn into a depression."

Fujita had found a new foe. But it would always be the tornado with which he was most closely associated. His work also had captivated the press and the public. He was always eager to turn on his tornado machine for the television cameras. He would as quickly accept an invitation to speak to school students as to a meteorology convention.

He became meteorology's best public relations man, a showman who, despite his stilted English, could explain a thunderstorm and a tornado to a packed room and leave the audience both enlightened and amused. And he loved the publicity.

In the summer of 1982, Fujita found himself west of Denver, where the high plains tilted toward the base of the Rocky Mountains. Fujita was tracking his new prey, the downburst. He ordered the three small Doppler radars, which were attached to flatbed trailers, to point westward. The downburst, a tornado's atmospheric cousin, was an extremely rapid sinking of cold air that Fujita believed was silently claiming hundreds of lives.

It was late spring, and the storm clouds began to form in the heat of the afternoon. The clouds spun rapidly into a thunderhead, and Fujita slipped out of the truck with his camera. Far in the distance were two tornadoes, both appearing a cool white from the rain condensation. He snapped the shutter like an excited schoolkid. There were cheers all around.

At age sixty-one, an elated Fujita saw his first tornado.

The JAWS (Joint Airport Weather Studies) research crew held a big celebration at the local Farmer's Daughter Restaurant. They ordered a cake decorated with a little chocolate twister surrounded by blue icing script: "Mr. Tornado Sees His First on 6-12-82." There were toasts all around.

Those would be the only twisters he would see. Fujita was never a storm chaser. He lacked the patience to sit at some truck stop on the Great Plains waiting for a thunderhead to develop. He preferred the view from the Cessna, to look down upon the mass of devastation below him from the storm's point of view. This was how, in his mind, he could envision the invisible winds, how he could make order from chaos. And a tornado was all about chaos.

10

A Twister's Journey

THIS IS THE FIRST SIGHT JORDAN remembers: a disemboweled horse lying on the road, a piece of lumber protruding from its ribs, and Chad Erwin standing over it, whacking it in the head with a trailer hitch. The image seared itself into the six-year-old's brain. Chad thought he was being humane, ending the suffering of the barely breathing mare. For Jordan, it became a mental snapshot that never faded.

A few moments earlier, Chad had tried to coax his family out of the culvert, but they wouldn't budge. He wanted them to see the black curtain as it dragged across Southern Hills on its way west.

"How's the house?" hollered his wife, Kitty.

"What house?" responded Chad.

"That's not funny."

"I'm not joking." It was gone.

"I mean, it was a cold feeling, man; I felt the blood just drop through my feet when I looked over here and saw this leveled down. I mean, everything around was gone. I mean, it was like—*The Twilight Zone*, they made a picture one time where an old boy came out of the cellar after a bomb went off, and that's exactly what I felt like, was that guy in that movie. I felt just exactly like him. There wasn't nothing around."

The tornado entered Bridge Creek through the Willow Lake addition, another mobile home subdivision, swept two frame-built houses off their slab foundations, and hurled a dozen cars more than a quarter mile before reaching Southern Hills, where Kara Wiese lived.

"All mobile homes in this area in the direct path of the tornado were obliterated, resulting in a high concentration of fatalities," an NWS survey found. The F5 powerhouse peeled back a one-inch layer of asphalt along one paved road. Its girth through Bridge Creek was more than one mile. It destroyed more than 200 mobile homes and site-built houses. Twelve people in the community of 1,500 died.

From the ridge, Chad could see its path. It looked as if someone had taken a wire brush and scoured the countryside. The tornado not only destroyed homes; it devoured trees, pulled grass up by its roots, and left a trail that would be visible for years. An electrician's van from a town twenty miles away came to rest in Chad's yard. It was melded to a passenger car. Kara's 1996 black Monte Carlo was on Chad's side of the ravine; a piece of lumber had rammed the windshield and embedded in the driver's seat.

As Chad walked down the road, he found the dying horse and then turned and saw a little boy. Jordan was walking toward him. The youngster was bloody and bruised.

"I asked him where his momma was, and he said they had been trying to get into the bathroom of their house. He said he felt her being yanked out of his arms." Jordan complained that his shoulder hurt.

Chad's wife took Jordan to the Bridge Creek School less than a mile away. The school was the only landmark left standing. By default, it became the central meeting place for the injured and homeless survivors.

Grady County deputy Robert Jolley drove a safe distance behind the twister and watched it go through Bridge Creek. He also had been trying to keep the roads clear, stopping to move large tree branches and scrap metal. He saw Robert Williams, in shock, staggering down the road.

"My wife's dead," Williams told him. "I can't find my daughter and grandbaby."

Amy Crago, his daughter, had been hurled nearly 100 yards. A neighbor found her in the grass and helped her into the back of a pickup that took her to the school. Ambulances from Chickasha were starting to arrive. Amy caught one of the first rides to the Chickasha hospital. She thought her entire family was dead.

Jolley drove to the lot where the Williams's trailer had been. Tornado trash was everywhere. Jolley began walking a tree line near a ravine, kicking debris away as he went.

At the base of a tree, he thought he saw a rag doll, rolled up in mud like dough. He grabbed it by the leg. Ten-month-old Aleah let out a squeal. The camera on Jolley's cruiser captured the scene that played repeatedly on local television. He sat the mud-covered Aleah on the hood of his car and checked her for injuries. The tearful baby reached out to him. "That's when I knew we would be okay," he said later.

The National Weather Service would later retrace the path of the tornado. A9 followed a nearly straight path paralleling Interstate 44 but for one exception: it made a slight jog and enveloped the underpass where Kathleen Walton was pulled to her death.

A passerby held tight to a compress on Pittman's bleeding leg. Another man had a severe fracture. Both were slathered in mud. A passerby hugged Levi as he cried for his mother. A state trooper found an ambulance to take them to the hospital, and Levi held Scott's hand for the entire trip.

Rather than giving shelter, the highway underpasses actually act as wind tunnels, increasing the wind speed. It was even more dangerous than if they had been flat on the ground. It took searchers an hour to find Kathleen Walton. She was in a nearby field, her body torn by the winds.

BY THE TIME TERRY Brown arrived at his Bridge Creek School superintendent's office, a small classroom trailer already had been turned into a morgue. The scene was chaotic as people were running around the buildings and pickups filled with the injured arrived one after the other. Volunteers and a lone paramedic, Steve Finley, treated the wounded brought to the school's gymnasium. Emergency management wasn't part of Brown's training as a school superintendent, but there was no one in charge. The former basketball coach took command.

"You want to get the morgue away from the road. We need a larger area, and you want it on an area that's going to be easy to clean when we're finished." Terry directed them to the elementary school cafeteria on the back side of the school.

Volunteers began moving the bodies, bodies they had wrapped in carpet or blankets because their clothes had been stripped away. Two victims had been employees of the school. Brown didn't recognize them.

"Literally the dirt was sandblasted into people. Even though you knew these people, you couldn't recognize them because of the way the mud was sandblasted into the flesh. They just didn't look like anyone you knew."

Several school workers who lived in Bridge Creek also showed up to help. The federal programs director told Brown she felt she had been called by the Lord to look after the bodies. Kathryn Turner gently covered each body with kitchen aprons. Among the first morgue victims were an older woman and a three-week-old baby, the Darnells. Kathryn held the baby, rocking him back and forth.

Paramedic Steve Finley arrived as the gymnasium was filling with the wounded. The sharp-featured Finley, his long hair pulled back in a ponytail, had been in an Oklahoma City ambulance diverted to Bridge Creek after a "hysterical distress call" from a volunteer fireman. The ambulance, unable to find the community, had to pick up a volunteer to help direct them to the school. Finley had no idea what he was walking into. Two hours passed before more paramedics arrived.

Finley moved from one person to another. "Whaddaya got?" Finley asked repeatedly as one severely injured person after another arrived. "Where do you hurt?" Emergency Medical Services Administration (EMSA) paramedics had stuffed bandages, splints, and supplies onto each ambulance and dispersed around south Oklahoma City to be ready for the twister's aftermath. He could not allow the lone ambulance to leave, and he could not find any transport help.

"My greatest fear was that the gymnasium, instead of a triage center, was going to become a morgue."

Sweat beaded across Finley's forehead in the sweltering, dark gym.

They had no electricity, no water, few supplies. School secretary Janie Pruett and other Bridge Creek survivors held the hands of their neighbors, trying to clean their wounds and comfort them. Bridge Creek's volunteer firefighters brought a steady stream of injured. Finley moved from one critical case to another, repeatedly trying to call for help on his cell phone. All the circuits were busy.

"At one point, they were coming in so fast, it was really chaotic and almost impossible to keep things sorted out," Finley said. His only source of light was from a television camera. The Discovery Channel show *Paramedic* had just happened to choose Oklahoma City EMSA to spotlight that week. "At the forefront of my mind was the worry that we've got to get some help here. We've got to get some people who can transport some of these patients out. There was always a question of how many more were coming and how bad would they be," Finley said.

Tom Tinneman's two daughters—Tawny, age nine, and Kylee, age six—were in critical condition with head injuries. Tawny especially was unresponsive. Tinneman himself had a fractured pelvis, a severed tailbone, and a broken collarbone. More Darnell family members arrived. Firefighters found three-year-old Wyatt in a tree. Blaine Darnell, Wyatt's father, vomited blood.

"What's this? Where is all this blood coming from?" Finley asked repeatedly as he ran to Blaine Darnell's side.

Finley placed the Tinneman sisters in the back of a police car. Ambulances from Chickasha and the surrounding area began to arrive, lining up beside the gym. Jordan was among the first to go.

"Gosh, how to describe it. It was just the worst, just the worst," Finley said.

As soon as the phone clicked dead, Mary Wiese called Dustin. "Don't worry, don't worry," Dustin told his mother. "I'll go out and check on her." Dustin, Misty, and two of their friends piled into the car for a long drive through the back roads to get to Bridge Creek. It was nearly dark by the time they arrived.

As they drove on the dirt road, near Bridge Creek School, they saw an overturned car. But a mobile home on the other side of the road seemed fine. "We thought just the phone lines had been knocked out and that's why we couldn't get ahold of her," said Dustin. He drove around the ninety-degree curve. "There was just nothing left." No crape myrtles, no mobile home, no car, no Kara or Jordan. "It was such a huge shock. On one side of the ravine, there were damaged houses, but they were still there. It was like it started at Kara's," said Misty. Dustin ran from the car. Chad told Dustin that Jordan had been taken to Bridge Creek School, but he didn't know anything about Kara. Misty drove to the gymnasium; Dustin waded into the muddy field looking for his sister.

Mary watched the television reports of the destruction. Dustin hadn't called her back. Unable to wait, Mary drove toward Bridge Creek. When she got to the dirt road leading to Kara's, she felt an enormous relief. From her car's headlights, she could see houses and mobile homes on either side of the road were fine. Everything was intact; everything was where it should be, right up to the curve in the road. And then her heart sank.

Two men walking by told her that all the injured had been taken to the school. She didn't see Dustin searching in the darkness. He searched frantically through the crushed cars that littered the area and the piles of debris. Mary did find Misty, who told her Jordan had been taken to a hospital.

Throughout the neighborhoods in the southern metro area, people emerged from their storm cellars and basements. Thousands of people had heeded the tornado warnings. Neighbors squeezed thirty or more people into a space made for half that or sought out the few home owners who did have basements. Some simply drove away, out of its path. Many hid in their closets and bathrooms.

The intensity of A9 had decreased dramatically but briefly as it crossed the South Canadian River into southwest Oklahoma City. Its power dropped to an F2, but it regrouped quickly. It damaged fifty homes at Country Place Estates, a dozen of them had F4 damage. One car was found under a bridge three miles away, deposited there when the tornado regained its F5 power. Police officers found an airplane wing—the wing from the Chickasha airport had been carried more than thirty miles.

At full strength again, the tornado had entered Eastlake Estates, a densely packed townhouse subdivision, killing three people and leaving nothing behind but slab foundations.

Across the street from Eastlake, the tornado flattened portions of Emerald Springs Apartments, a 600-unit building. Police and firefighters worked through the night pulling the apartment dwellers from beneath the rubble. They also found three bodies. "It looked like a war zone," said Jim Primble, a resident of the apartment complex. "It was the Oklahoma City bombing all over again." Oklahoma City firefighters spray-painted a large red X over each pile of searched rubble as they methodically worked their way through the neighborhoods.

North of Eastlake Estates was Westmoore High School. Remarkably, none of the parents or students attending the awards ceremony was injured, but the twister sprayed some of the 200 cars in the parking lot around the neighborhood. A dead gray horse rested on its side between cars.

John Graham, a six-foot-five EMSA paramedic, and three crew members established a triage center across the street from Westmoore High School. The walking wounded came first—those who suffered cuts from flying shards or bruises but were able to get to Graham on their own.

After the first wave, Graham hopped aboard a fire department brush pumper, normally used for grass fires, and rode with the firemen into the additions. They joined in the house-to-house searches.

"You could see how people survived. The house would be totally destroyed, but just the closet would be standing. You could tell people had been hiding in there by the bedding they were under. Ten houses would be destroyed; then there would be one house with absolutely nothing wrong with it. The roof was lifted off of one house and then set down perfectly on top of a neighbor's house. And nothing had moved inside the house with the missing roof. Then you would see cars that were crushed into little pieces."

Searchers also found two bodies.

The crushing, cutting injuries Graham saw were not unlike those from the 1995 bombing of the Oklahoma City federal building. "It didn't have the direct impact that the bomb did, but you do have a lot of flying debris that injure people." Most of the people killed across the street from the Murrah Federal Building had died from flying glass and metal.

As the twister entered Moore city limits, it again grew more powerful, producing F5 damage three-quarters of a mile wide. Luckily for Julie Rakestraw, her neighbor did not touch the splintered lumber jutting from her head. He wrapped it in gauze to keep it stable. She was one of the first to arrive at the hospital.

Stunned doctors at Hillcrest Hospital pondered how to remove it. They searched the hospital's maintenance department until they found a circular saw. They sterilized the blade and cut away the lumber. Rushed to the operating room, Julie underwent an arduous surgery to remove the piece of two-by-four. The impalement caused permanent nerve damage to the right side of her face, but she lived.

All hospitals went Code Black, signaling a major regional emergency. Off-duty personnel were summoned to help. Ambulances driven by professionals and cars and pickups driven by citizens backed up in front of the emergency room doors, unloaded the wounded, and headed back to the devastation. As the night wore on, frantic relatives paid repeated visits to hospitals around the city, searching for missing kin. Telephone service had become sporadic, and cell phone circuits jammed.

In the hardest-hit neighborhoods, survivors ran from one ruin to another. "Is anyone here?" they shouted. The digging was done by hand, by neighbors.

A9 produced its greatest damage in Moore on Janeway Avenue, Charlie Cusack's neighborhood. Four people were killed as blocks of homes were flattened.

All the ordinary landmarks—street signs, neighbors' houses—were destroyed. Charlie Cusack felt disoriented when he emerged from the rubble that was his home. There was nothing left that gave a sense of place or direction. The one-ton flatbed truck he had driven home was in the middle of the street. Beside it was a small Toyota, wheels up. Someone's motor boat had come to rest atop the Toyota and a tree had fallen on both.

"It looked like an atomic bomb had gone off," Charlie said.

His neighbors, Lloyd and Ann, came running out of their destroyed home. Seconds later, the street vibrated as an explosion ripped up through the debris at the end of the block. A natural gas line was on fire. The two families started walking in the opposite direction. At the next intersection, Charlie saw his brother Al running toward him.

A couple of Charlie's ten siblings had been at his parents' house five miles away. They all had stood in their parents' front yard and watched the twister make its way through Moore.

"Somebody said, 'Now, Charlie and Pam are over there; that's over by their house.' And they're like, 'Nah, they're okay.' And my sister-in-law Robin kept saying, 'No, somebody needs to go over there and get them.' And so my brother Al got in his truck and drove over there." What started as a grudging errand for his wife turned into a chest-thumping search as Al inched closer to the devastation and trash in the road kept him from driving farther. "He was frantic. He didn't know where he was and he had been to my home many times. He was just running down the street. I think he would have kept going if he hadn't seen us."

The Cusacks loaded into Al's truck and drove to Charlie's parents' house. It was easy to move back in; they had nothing. "There are eleven of us, so just about everyone has moved back home one time or another."

The tornado's intensity fluctuated dramatically, its winds slowing after it crossed I-35, and caused minor damage to the First Baptist Church of Moore. The church later served as the center for tornado relief efforts. A9 powered up as it moved east of I-35 and entered an industrial park area and a lightly populated section of the city. It destroyed two major businesses, including American Freightways, where it killed the two truckers. It also picked up an eighteen-ton freight train car and dragged it for three-quarters of a mile through a pasture. "Gouge marks were observed in the field every 50 to 100 yards, suggesting the freight car had been airborne for at least a short distance," the NWS survey noted.

In Del City, Carolyn Stager remained awake throughout the entire ordeal. There were people in uniform standing over her now. Military personnel from nearby Tinker Air Force Base had flooded the Del City neighborhood. Carolyn's pelvis was broken, her head bleeding. Whatever hit her cleanly peeled back her hair.

"It had to be something like a piece of paper or a napkin. If it had been something sharp, it would have killed me."

They bandaged her head. The airmen turned doors and pieces of plywood into makeshift gurneys, using anything they could find to help carry the wounded to a triage center on the base.

Her pregnant daughter had been battered by the flying bricks and nails and glass. Something tore out the back part of her leg. "She looked like someone had taken a baseball bat and beat her everywhere and then taken a shrapnel gun and sprayed everywhere but her face and her stomach, which was amazing."

Tinker medics put Carolyn on the last ambulance to St. Anthony's Hospital. Her daughter was sent to a hospital in Midwest City, a suburb north of Del City.

Six people, including two of Stager's neighbors, died in the suburb. Her Del Air neighborhood, she was told, was hit with the tornado's full force even though it was much narrower there than it had been at Bridge Creek. Nothing of Carolyn's house remained. Nearby homes at least had walls standing or closets in which her neighbors had sought shelter.

One of the first reporters to the Del Air neighborhood was Jack Damrill, the RedHawks play-by-play man who also worked part time for Channel 4. He grabbed a Channel 4 intern who had been photographing the RedHawks game, and the two headed east. The ten-minute drive to Del City took him ninety minutes because of all the traffic. "It

was incredible just watching people walk around, in shock, not know-
ing what to do. That subdivision was gone. There was not one house left
standing."

The Weather Service survey found A9 had sideswiped a main entry
gate to Tinker Air Force Base before moving northward into Midwest
City, where it plowed along the Interstate 40 corridor, an area of strip
malls, car dealerships, restaurants, and hotels. It damaged 800 new cars
at Hudiburg Auto Group, lofting dozens of them across the interstate
highway. It destroyed the motels, the Cracker Barrel restaurant, and part
of Rose State College, where Kara had gone to school.

Rescuers picked through the debris-strewn parking lot at the damaged hotels, writing down license tag numbers from damaged cars so
they could try to track down the owners. National Guardsmen dug by
hand through the rubble of homes behind the hotels, finding the body
of one young boy.

In a last gasp, A9 entered one more neighborhood north of I-40,
killing three people. And then it disappeared. The average time it took
to destroy one house: twenty-five seconds.

The destruction was so massive and so widespread that the initial
fear was of hundreds of fatalities. In Moore, the fire chief surveyed the
scene and called city officials: "We'll be lucky if we don't have 400 to
500 people dead." Temporary morgues were set up around the area: the
choir room at the First Baptist Church in Moore, office building lobbies
in Del City, downtown stores in Midwest City.

As THE TORNADO ENTERED the metro area, the rear flank downdraft,
the torrent of wind and water, was so heavy that Craig Wolter lost
sight of the DOW truck. The school vans also were getting hard to
handle in the strong gusts that slammed the vans even a mile from
the twister.

"It's hard to visualize this, but there's the tornado straight ahead, and
then there's debris and all the stuff that goes up out of the tornado and
kind of creates a curtain behind you, so that no matter which direction

you look, it looks black. And it was getting to that point, where I just said, you know, 'We're going to stop here and let the storms kind of go ahead of us.'

"You couldn't have asked for a more textbook way for a science teacher to show the mechanics of a tornado. It was just perfect. It was parallel to us along the interstate," he said. The vans stopped and A9 crossed the South Canadian River into southwest Oklahoma City and Moore.

The Minnesotans drove through side streets in Moore trying to find a gas station. The vans were running on fumes. That's when they saw the other side of the tornado, the aftermath. From a gas station, the students saw the twister's victims beginning to emerge from their homes at the edge of the damage path. They saw a triage center being established by EMTs and asked if they could help. They were told to clear the area.

It was a quiet and somber ride back to their motel. As they arrived, the parking lot already was starting to fill with the newly homeless victims, some driving cars that, though still running, were dented and had broken windshields. Wolter told the kids to clean up, and they would regroup for a discussion. While they were in the parking lot, a couple emerged from a beat-up car.

"Were you guys out chasing with these tornados?" the man asked.

A lump rose in Wolter's throat. He was reluctant and embarrassed to tell the man that, yes, they had been out chasing the tornado. "I feel uncomfortable telling you about our experience, since I see obviously you lost your home or whatever and have some hardship here."

"Hey, you know what, in Oklahoma you have tornadoes. That's what happens. It's up to you guys to research them so that people like myself are right now walking instead of being dead."

Wolter and the students did not realize the full magnitude of the disaster until they turned on the television. They stayed up until dawn, watching the news and talking about their experience.

"Every time I said, 'All right, guys, let's turn in,' the kids would bring up some other scientific thing about it. 'Why did this happen or why did that happen?'"

Roger and Rich watched the biggest tornado of their storm-chasing career disappear into the night, aiming for Kansas. Roger turned the Meatwagon south, and they headed back to Norman. Roger had to be at work at midnight. And he guessed, rightly, that the Storm Prediction Center would be busy throughout the graveyard shift.

They swung around the west side of Oklahoma City, trying to avoid the traffic jams and blocked roads and broken homes they had been hearing about on the radio.

They could not see the destruction around them, but they could smell it. It was the sweet aroma of cut grass, and it grew stronger. There was the fresh smell of vegetation ripped from the ground, of trees stripped of their bark, and of newly cut lumber. It was the scent of the pine trees that Charlie Cusack smelled wafting under his bathroom door. There was mustiness in the air from the dirt and wet insulation. And there was the pungent smell of natural gas.

"It was one of the most unforgettable sensations in my entire life—much less my history of storm observing," Roger said. "The unique and pungent aroma of natural gas, combined with shredded vegetation, became very strong. We realized we were smelling the scent of devastation, carried on warm southeast breezes for ten to fifteen miles from the Bridge Creek/Moore tornado's swath."

The odor lingered for days.

BRIDGE CREEK SCHOOL SECRETARY Janie Pruett steeled herself with a prayer and walked into the makeshift morgue. "I cannot stand the smell of blood. But I just prayed when I went in there and told the Lord if this is what you want me to do, I'll do it. He took that smell away. I couldn't smell the blood all night long. I couldn't even stand to look at anyone bleeding without it making me sick. The Lord gave me the ability to do what I needed to do."

The bodies kept arriving a few at a time all night long. They were wrapped in comforters, sheets, afghans, anything searchers found lying around that gave them some decency. "We couldn't identify any of the bodies," Janie recalled. Janie; her husband, Dennis; and Kathryn Turner stayed in the morgue all night, keeping vigil over the dead.

People were still looking for Kara Wiese. An elementary school secretary who knew both Jordan and Kara tried to identify bodies in the morgue. "Kara was there, but we couldn't be for sure it was her." She had been found in the ravine.

The bodies whose identities they knew for sure were the three-week-old baby and his grandmother, Ashton and Lucille Darnell. The baby was wrapped in a sheet and placed next to his grandmother.

By 2:00 a.m., when the medical examiner arrived, there were eleven bodies in the school cafeteria. One funeral hearse after another arrived as the drivers placed the bodies in bags. They gently placed Ashton on his grandmother's chest for their final ride together. Kathleen Walton's torso was placed in a body bag.

Every neighbor who came by made a mental roll call of the Bridge Creek community. "What about so and so?" Janie remembers people asking over and over again. She had lived in Bridge Creek for twenty-five years, had played the keyboard at Ridgecrest Baptist Church for twenty years, and had been school secretary that long as well. She knew just about everyone in the community, and with each person she saw walking, alive, she made another mental checkmark. "When you saw someone you knew, it was just a joyous occasion."

At sunup, the sounds of pistol and rifle shots echoed across Bridge Creek's ridges and ravines. Deputies and owners shot wounded and dying cattle and horses; one policeman shot a horse running wildly and blindly in a pasture, its eyes sucked out by the twister.

It was nearly dawn before the Pruetts discovered that they still had a house. They had left once the tornado warning was issued and after Janie heard all the troopers' sirens along I-44. She and her husband drove into Oklahoma City, seeking shelter at a convention center that her husband was helping to renovate. Their home was still there; a downed tree rested on its roof, but the house was undamaged.

"You were anxious to know if you had a house, but if you had a house, you felt guilty because your neighbors didn't. If you lost your house but you still had stuff around, then you felt guilty because you had stuff. If you lost everything, you still felt guilty because you were alive. There was so much guilt. You felt guilty no matter what," said Janie.

OKLAHOMA CITY OPENED ALL of its schools to help shelter the thousands of homeless survivors. Churches did the same, and the huge First Baptist Church of Moore became both shelter and command center. Hundreds of mud-covered, wet survivors walked to the church, some carrying a few clothes they had scavenged or anything else that remained whole. One person brought a fishing pole; another carried a birdcage.

On Tuesday morning, survivors awoke to little relief. There were threats of another tornado, and the skies darkened once more. It rained; it hailed; but there was no twister, just a reminder of the previous evening. Rescuers went back to work and, with trained dogs, continued pulling survivors from ruins. They found one father and son in such shock they refused to come out of the rubble. Several families were trapped in their own storm shelters by piles of debris.

More than 140,000 homes were still without electricity. Power losses caused raw sewage to flow into the South Canadian River. There were problems with drinking water and sewer lines. A9 had fractured the region's entire infrastructure.

Restaurants around the region provided free food to the rescue workers and the survivors. Hotels offered free rooms; the University of Oklahoma cleared some of its dorms. Wholesalers brought tens of thousands of paper goods and semitrucks filled with goods. After only a couple of days, recovery organizations asked people to stop bringing food and supplies.

The cleanup started immediately. It also meant carting away vital clues.

The Federal Emergency Management Agency (FEMA) hurriedly put together damage survey teams. They wanted to study it from the ground and the air, just as Fujita used to do. They had to rush because trucks already were hauling away debris. FEMA appointed Chuck Doswell to its Building Performance Assessment Team.

Doswell didn't like doing damage surveys. He didn't like seeing people's personal stuff strewn about for the world to see. It reminded him too much of his own stuff. As Doswell walked through the ruins, he wondered what had happened to the people. The bricks and broken lumber didn't bother him. But when he saw a toy or some other personal item not intended for a stranger's eye, he wondered, *Did they make it?*

"The analogy with a 'war zone' is compelling. What was whole becomes junk in seconds, just as experienced by the victims of bombing raids . . . but the atmosphere's very indifference renders the whole thing surreal. At least with a bombing raid, the *intent* is understandable. Here, hubris and humility alike are swept away by the uncaring atmosphere. The tornado doesn't care about what it encounters. . . . it's not evil, just unfeeling," Doswell wrote in a log. "Does that make it easier or harder for the victims to cope? I don't know. I get the impression that many of the victims are searching for *meaning* in this disaster, but, from my intellectual perspective, they are simply unlucky. I also sense they don't want to hear that this event was just a roll of the dice."

11

VORTEX

WEATHER GEEKS PILED INTO THE UNIVERSITY of Oklahoma van and headed across the prairie, the cassette player cranked to the max as Vivaldi's *Gloria* screamed from the speakers. Howie Bluestein was in the driver's seat, part of a new generation of tornado researchers coming of age in the 1980s. Fujita preferred the silent, poststorm aerial surveys, but the new scientists wanted to be in the action, to be as close to the tornado as possible. At the National Severe Storms Laboratory (NSSL) and the University of Oklahoma, the chase was on, literally.

Fujita and his era provided science with the tornado's architectural framework, unlocking some mysteries, but the keys to the bear's cage remained hidden. Changing technology gave a tantalizing glimpse into tornado mechanics, but still did not provide a definitive answer.

Bluestein, bushy curls atop his head and a Boston accent on his lips, edged closer to the maelstrom. Bluestein had come to OU from MIT in the late 1970s and helped pioneer the scientific storm chase. On stormy days, he and his students raced across the countryside in search of a tornado. There were no cell phones, no laptops, no Global Positioning Satellites. If they saw a twister, they marked its location on a paper map and stopped at the first pay phone they saw to call the scientists at the storm lab. They helped NSSL do the legwork for the new Doppler prototype, calling in tornado sightings that lab researchers matched to Doppler scans.

Rather than being passive observers, scientists wanted a peek inside, but weren't sure how to get it. They wanted more information than just the winds made visible by the Doppler. They were like plumbers searching for just the right wrench, only they had to make the wrench from scratch.

In 1981, Bluestein and others created the Totable Tornado Observatory, a fifty-five-gallon hardened drum crammed with instruments. The idea of TOTO, named for Dorothy's dog from *The Wizard of Oz*, was for a chase team to put the 400-pound package in front of a twister and let the tornado run over it. It was easier said than done.

Once TOTO was loaded onto the back of a chase truck, researchers had to find a tornado, divine the exact direction of the tornado, get in front of the tornado, find graduate students strong enough to unload the heavy TOTO in thirty seconds, flip a switch, and, as Howie put it, make a strategic retreat. They got close, but it never happened. Howie tried for three years; NSSL tried for a couple of more.

The importance of the scrappy TOTO was not limited to science; it had greater financial value and success as a screenplay plot device. In 1996, TOTO morphed into DOROTHY, the 400-pound, 55-gallon drum that fictional scientists in the movie *Twister* tried to place in front of a tornado.

Scientists at Los Alamos National Laboratory had an even more inventive idea. They tried using an airplane to fire lightweight, instrumented missiles into a tornado. Unfortunately, the rockets had to be light enough not to kill people on the ground, which made them too flimsy to penetrate a tornado. Plus, the airplane almost crashed after getting caught in an updraft. Interest waned in that project, but only temporarily.

Another scientist tried placing sound instruments near the tornado to record its "sounded just like a freight train" noise. That didn't work either.

Howie and crew tried another idea: releasing weather balloons into a tornado. On the first attempt, the underinflated balloon scraped across a wheat field in the general direction of a tornado but never did quite make it. Undeterred, they tried again and became rather good at it. With their balloons, they were able to measure the incredible intensity of the updraft gusting into a supercell at fifty meters per second. They were able to study moisture and temperature levels.

In the late 1980s, OU researchers also created small instrument packets they called turtles because of their appearance. The idea was similar to TOTO but far easier to carry out. The lightweight turtles, their shells looking just like their namesake, were placed at numerous locations around a tornado to ensure data collection. Because they were low to the ground, they also often stayed in place as the winds grazed over their aerodynamic surface. From the turtle, scientists could learn the air pressure, humidity, and temperatures in and around the twister.

Researchers returned to the airplane idea, this time by attaching a Doppler radar to the plane. Hurricane researchers had successfully used two airplanes, each equipped with Doppler, to scan the same hurricane from different angles. A field experiment on supercells proved difficult. In the trials, the tornadoes dissipated as soon as the airplanes arrived. Tornadoes lacked the cooperative, long-lived nature of hurricanes.

The storm lab hit on the idea that a high-resolution scan of the tornado could be made if the Doppler were physically carried to the tornado. Bluestein and his students packed a small, portable Doppler in the truck and started searching for a supercell. It worked. It was still cumbersome, but now they had a tool they had been searching for.

Using the portable Doppler, Bluestein for the first time documented a violent tornado in 1991, a huge twister near Red Rock, Oklahoma. Doppler clocked it at 286 miles per hour. All previous powerful tornadoes were classified as such according to Fujita's estimates based on flying debris or the degree of damage. Bluestein's tornado proved the winds really could reach such levels. It was a record that would stand until May 3, 1999.

Josh Wurman took a look at Bluestein's project and thought, "Right idea; wrong toy." That's how he came up with the Doppler on Wheels, a self-contained truck that bristled with computer instruments and had the Doppler strapped to its bed. In the DOW, Josh could keep pace with the tornado, pause, take some pictures, and continue the chase. Wurman and Erik Rasmussen, in separate DOW trucks, eventually would corner a twister and obtain the first three-dimensional triangulation of a tornado that produced tons more data from different angles.

Trying a variation, Bluestein placed a powerful MIT Doppler in an OU van and, like a submarine periscope, raised and lowered the radar through a hatch. It worked fine—when it wasn't raining.

Despite the hit-or-miss remote-sensing ideas, researchers crafted new theories about tornadoes based on their discoveries. Bluestein studied the plain vanilla thunderstorms, the nonsupercells, that produced frequent but often weaker tornadoes—landspouts, as he called them. The nonsupercell thunderstorms lacked the distinctive rotating wall cloud of supercell storms. Their benign appearance and weaker radar signatures made it difficult for forecasters and for NEXRAD to detect the short-lived twisters. Bluestein found these types of funnels formed from intense surface winds stretching and ascending upward, where they were met by an intense rear-flank downdraft that wrapped around them.

The tornado did not "touch down" or descend from the cloud to the ground. The midair meeting of currents depended heavily on surface-level winds, temperature, and moisture. These subtle factors were almost impossible for an operational forecaster to detect beforehand. Questions also remained about the supercell and its violent offspring.

Researchers prepared for one major expedition. They rounded up their turtles, their DOWs, their vans, their planes, and all their other whirligigs for the world's biggest scientific hunt.

FOR ONCE, FUJITA WASN'T looking for a twister. He hunted for an entity that no one but he and a few others thought existed, an entity so dangerous that it was killing hundreds of people without being noticed. Some people thought Fujita was off on a harebrained idea. Airline pilots knew otherwise.

On June 24, 1975, a Flying Tiger Line DC-8 cargo plane nearly crashed at Runway 22L at New York's Kennedy Airport. A storm was brewing along the coastline. The pilot warned the tower of an extreme wind shift near the end of 22L and suggested it change runways. The tower reported it recorded winds of fifteen knots. "I don't care what you're indicating. I'm just telling you that there's such a wind shear on the final on that runway you should change it to the northwest," the pilot radioed back.

A minute later, an Eastern Airlines L-1011 jumbo jet encountered such severe shear that it came within twenty feet of clipping the approach light stanchions at the end of the runway. The pilot aborted, gunned the engines to full takeoff power, and flew to Newark. Another DC-8 and a small propeller plane landed without problems. Eight minutes after the Flying Tiger cargo pilot had recommended a runway change, Eastern Airlines Flight 66, a Boeing 727 flying from New Orleans to New York, was directed to 22L. "Okay, we'll let you know about the conditions," Flight 66's pilot told the tower.

A half-mile from the runway, Flight 66 fell into the steel light stanchions, bounced, and flipped upside down. There were 113 fatalities among the 124 passengers. At the time, it was the worst plane crash in U.S. history.

By 1975, if the issue dealt with storms, tornadoes, or winds, there was only one person to call: Ted Fujita. Eastern Airlines contacted Fujita for an independent study.

Fujita began his own analysis by reviewing the damage caused by small-scale winds during the 1974 Super Outbreak of tornadoes. Something about the aircraft warnings and the crash site pictures jogged his memory. Wind shear—extreme variations of wind direction and speed at different altitudes—did not seem to explain the situation, at least to Fujita.

Looking at the Super Outbreak aerial photographs, Fujita noticed the "mysterious starburst patterns" of uprooted trees. "Such a damage-causing downdraft might be extremely rare but it could happen, I thought." These were similar to the starburst patterns he had first witnessed at ground zero in Nagasaki and Hiroshima, the burst that left the trees directly beneath the atom bomb explosions still standing but felled others in rippling circles.

Vortex

He discussed the starburst pattern with Horace Byers, who suggested Fujita call it a "downburst" to convey both the downward and outward pattern. It would be Fujita's most elusive quarry. He interviewed witnesses to the crash. He reviewed the weather data for the region. He talked to pilots who had landed just before Flight 66. And, as always, he crafted an amazing chart of the flights into Kennedy. His theory was that there had been three downbursts, or microbursts, as he later called them, within an eighteen-minute period. The microbursts were too concentrated to be recorded by runway wind gauges.

He mapped the microburst wind directions and the effects on each of the eleven jets that landed before Eastern Flight 66. Some of the planes reported headwinds, some tailwinds, and some nothing at all. The Eastern flight, he surmised, flew into the heart of the third microburst. In the microburst, pilots first encountered unexpectedly strong headwinds, and just as they compensated, they experienced strong tailwinds, which robbed the wings of lift.

In a tornado, the winds curl and tighten into a funnel. The downburst had the power of an F2 tornado, but it was a straight wind, vertical from cloud to ground. Once it hit the surface, it bounced back into the air, curling like a puff of smoke, creating both a headwind and tailwind for the unfortunate plane entering its space. It was more than just a wind shear, Fujita surmised; it was an identifiable meteorological creature.

Pilots knew just what Fujita was talking about. Several called to thank him, reporting that they too had experienced unexplained wind shears over the years.

Many meteorologists, however, dismissed Fujita's downburst theory, suggesting he had confused a downburst with the well-known gust front caused by a brewing thunderstorm. Several suggested Fujita stuck a new name on an old concept.

"The criticism led to several sleepless nights. However, I did not want to abandon the downburst concept because of my belief and of increasing support from pilots who had experienced dangerous flights through downburst wind shear."

STORM WARNING

In his memoir, Fujita recalled the fate of Alfred Wegener, the German meteorologist who had made one of the greatest geological discoveries. Wegener's theory was that the global continents had once been attached, had broken apart, and had drifted to their present locations. Wegener came to his idea simply enough. By looking at maps of South America and West Africa, it seemed obvious they fit together like puzzle parts. Lines of ore, mountain ranges, and fossils all indicated to Wegener that the continents had been one landmass.

A meteorologist and astronomer by training, Wegener's entry into geology was dismissed as "utter, damn röt." Wegener's continental drift theory ultimately was proven correct, but not until three decades after his death in 1930. Fujita did not want his downburst theory to meet the same fate.

Fujita sought to prove his concept using new Doppler radar. He enlisted the support of the National Center for Atmospheric Research (NCAR) at Boulder, Colorado, which also had a Doppler radar. An associate suggested a project name: Northern Illinois Meteorological Research on Downburst, or NIMROD, the biblical name of a mighty hunter. The prey was the downburst.

Fujita and the NCAR team set up the Doppler radar outside Yorkville, Illinois. NCAR's Jim Wilson briefly explained the Doppler radar data to Fujita. He was stunned when Fujita could understand all the Doppler data and interpretations after a few minutes. "A scientific genius," Wilson called him. On May 29, 1978, they set up the radar and aimed it toward a flash of lightning. A storm was headed their way. Fujita and Wilson stood outside looking at the storm they could see on the Doppler.

A sudden, powerful gust of wind knocked them off their feet, almost blowing them into a farm pond. Fujita dubbed it a microburst, and the name would stick. "With great excitement they realized they had observed the first microburst on radar and had actually felt the diverging outflow," Wilson later wrote.

Still, the downburst theory continued to divide the meteorological community "into bursters and antibursters." Some critics suggested Fujita's insistence on a downburst was confusing and dangerous for airline pilots.

The pilots, however, remained on Fujita's side.

In 1982, a Pan American 727 taxied down a New Orleans airport runway in a heavy rain but reportedly calm winds. It barely left the ground. It crashed into a tree and slid into a neighborhood of homes. All 146 people aboard and 8 people on the ground died.

Fujita analyzed the accident scene and determined a microburst hit near the center of the runway and spread out quickly, preventing the jet from gaining altitude. The second wind-related crash prompted a national review, which made recommendations about wind shear but never mentioned the words *downburst* or *microburst*.

Fujita conducted his own study, which he released in 1984. In it, he suggested that at least eight major airline accidents that had claimed 514 lives since 1964 could be attributed to downbursts. Fujita even documented the most extreme case, a 149-mile-per-hour downburst that occurred at Andrews Air Force Base in Maryland only six minutes after Air Force One landed with President Reagan aboard. The wind shear tore the roofs from several hangars and blew trees across the highways. He urged that Doppler radar immediately be installed near Andrews, Air Force One's home base.

In 1985, Delta Flight 191, a jumbo L-1011, crashed on approach to the Dallas–Fort Worth Airport. Again, a storm had formed along the glide path. Fujita again analyzed the aircraft and the anemometers surrounding the runways. He found the jet had crashed within sixty seconds of entering a three-mile microburst at the end of the runway. By the time the anemometers recorded the shear, the jet was in flames.

Finally, the microburst supporters gained momentum. Said one meteorologist: "Myself and others have felt, since the late 1970s, that there were factors of embarrassment, frustration and professional jealousy in some of the early criticism. Here was a phenomenon that had literally hit prominent meteorologists in the face for a lifetime and they failed to recognize it as a unique, discrete phenomenon."

The Federal Aviation Administration ordered the fifty largest airports to install a low-level Doppler radar system that could detect the microbursts and ordered airlines to install downburst detection devices on airplanes. The airlines also began instructing pilots on ways to fly through a downburst.

The New York Times quoted one unnamed Reagan budget office official who was balking about the cost of airport Doppler radars as saying, "More people died last week in motorcycle accidents. How safe is safe?"

Fujita was too impatient to complete the entire theoretical details of many of his discoveries and too frustrated by meteorology's peer review process to officially publish many of his theories. It would be up to others to try to determine the physical manifestations of the downburst.

"However, the microburst was another example in his illustrious career where he stimulated an entire research community to focus their efforts on an important research problem," wrote Wilson and Roger Wakimoto, one of Fujita's graduate students. "In a period of only about 15 years the scientific understanding of the microburst evolved from no knowledge to a thorough understanding of the evolution of the downdraft and outflow and considerable knowledge of the forcing mechanisms of the downdraft."

FOR FIVE DECADES, SCIENTISTS had probed the thunderstorm and its offspring in fits and starts, at the mercy not only of rare appearances by reluctant twisters but money for the research. Finally, after months of planning and money in hand from NOAA and the National Science Foundation, Erik Rasmussen and a team of the most prominent tornado researchers in the country were ready with VORTEX—Verification of the Origins of Rotation in Tornadoes Experiment.

Not since Horace Byers's Thunderstorm Project in 1945 had there been such a comprehensive, all-encompassing scientific effort to understand tornadoes. It was the biggest tornado expedition in history, a far cry from Fujita's lone Cessna. VORTEX built on everything that had gone before it. More than seventy-five scientists from the United States and Canada, twenty cars outfitted with weather devices, three airplanes, and a hundred volunteers, radar operators, and chasers were organized into teams. By 1994, the two-year project was ready to go. The goal: tornadogenesis.

What research weapons they lacked, they invented. The University of Oklahoma scientists outfitted a dozen Ford Escorts with a mobile Mesonet—an array of instruments to measure the winds, temperature, and humidity and record data every few seconds—the "geekmobile," the student volunteers called it. Once researchers had to find a pay phone so they could report tornado sightings to the Weather Service; now there were laptop computers, GPS connections, and mobile phones. There were two airplanes, including an NOAA P-3 carrying airborne Doppler radar. Josh Wurman was there with his DOW trucks.

The idea of the tornado safari was to surround the tornado with the mobile Mesonets, weather balloons, and radar and collect as much data as possible. Rasmussen was the field general, dispatching the teams to their positions.

Everything was ready but the tornadoes; 1994 nearly set a record for the lowest number of twisters. VORTEX captured just one. The team did better the second year, corraling nine tornadoes, including an F4 at Dimmitt, Texas, that became the focal point of the research project.

Still, VORTEX did better at refuting hypotheses than confirming them.

For years, people had suspected the supercell's mesocyclone—the distinctive rotating updraft and downdraft—played a major role in tornado formation. It made sense, visually. The miles-wide mesocyclone swirled horizontally and counterclockwise. It was not such a leap to assume that the rotating winds in the mesocyclone could create a vertically spinning, counterclockwise rotating funnel that would descend from the cloud.

But as with all other things meteorological, that theory would be too simple, and VORTEX found it to be wrong.

The tornado did not form from the supercell's mesocyclone; it formed behind it. All the action was in the rear flank downdraft, that cool waterfall of air first discovered by Byers in the Thunderstorm Project and Fujita on the mountain ridge in Japan. And the formation of a tornado may have much more to do with what is happening at the surface than what is happening in the sky. Air is, like water, a fluid. Because a storm is a small low-pressure pocket, the ground air rushes toward it, and like a river's rapids, it tumbles and swirls and eddies on its journey. The creation of the tornado may occur because of what is happening just a few yards above the ground, similar to the landspout from weak thunderstorms.

"In VORTEX, we wanted to figure out how tornadoes form in supercells, and we didn't quite get there," Rasmussen said. "But we accomplished more than we initially thought. In the broadest sense, VORTEX increased our focus on what is happening right at the back of the storm."

In studying the data from the 1995 Dimmitt tornado, scientists found the rear downdraft split as it slid down the storm's back side; one side rotated clockwise and the other counterclockwise. What VORTEX researchers found in a follow-up project was that air temperature may be a key. The ground-level air must be warm and humid, and the downdraft must be cool but not too cold. The downdraft, cooled by evaporating rain, may wrap around the warmer air near the surface and rise again with the updraft. The two twines contract into a funnel.

Rasmussen believed there may be other features, such as raindrops, within the storm, within the downdraft, that would help forecasters better determine if a tornado was likely and whether it would be weak or strong.

Radar researchers in Norman already were working on refinements to the Doppler that would enhance its ability not only to see the winds but to determine if the raindrops were large or small, which could help forecasters determine the temperature of the downdraft and the possibility of tornado creation. With dual polarization, the radar can distinguish between rain and snow, provide forecasters with a clearer picture of the storm, and better detect the telltale signs of tornado formation. Dual polarization radar will be an interim step to the next generation of radar, called phased array. The phased-array radar is an offshoot of the military's SPY-1 radar used by Navy warships. Doppler provides forecasters with a picture once every six minutes. The phased-array radar will provide a snapshot once a minute. All of this means many more thousands of additional pieces of information to feed into the computer and create improved numerical models.

VORTEX

The phased-array radar is probably ten or fifteen years away from operational service—unless the government learned nothing from its twenty-year procurement ordeal for the Doppler.

It had taken fifty years, but scientists were tantalizingly close to unraveling the storm's mysteries. The clues were beginning to fall into place. All it might take would be one more giant safari, VORTEX2.

For millennia, people believed the tornado, the great wind, was a capricious act of God, that the tornado descended from the heavens. But it was an optical illusion: the twister spins up rather than down. Condensation from the storm makes the wind coils visible first near the parent cloud, making the funnel appear to descend. But the answer to the tornado riddle may be near the ground, not the sky.

12

THE TWISTER'S AFTERMATH

WHEN JORDAN, WHO WAS BRUISED, SCRATCHED, and had a cracked rib, awoke at the hospital, he knew of only one reason his mother was missing. It wasn't the mobile home or the 300-mile-per-hour tornado. It was his fault.

"I should have held on tighter," he told his grandmother.

Their trailer teetered from the wind, Jordan told Mary. He and his mother had been facing each other. Kara was holding him, and he was holding her. And then they were airborne, flying through the air. He saw a flying cow, he said. He saw a flying horse. And something started hitting him in the back. He felt his mother slip from his grasp, and then something heavy hit him, and he was knocked out.

Jordan was released from the hospital the next day. Dustin had returned to Kara's place to continue his search, walking through the trash-strewn fields. Mary had spent the night calling hospitals, hoping that Kara was lying unconscious in an emergency room. The alternative was too much to imagine.

Mary took Jordan to her home in El Reno. He fell asleep on the couch. Mary took pictures of his scratched-up legs so she could later show Kara what his injuries looked like. Kara would want to know.

Wednesday morning, the Oklahoma County coroner called Mary. Did she, he asked, have Kara's fingerprints by chance? When Dustin and

Kara were kids, the issue of missing children had become national news. Grocery stores, churches, and television stations offered free fingerprinting. Mary took them repeatedly. Not that she was worried. Dustin remembers that as a kid he thought it was fun to be fingerprinted, so he and Kara asked to go. Mary kept all the fingerprints in a shoebox in the back of a closet.

When the coroner called, Mary knew her daughter was dead. None of the hospitals had unidentified patients. And had Kara been hospitalized, she would have been identified by now. It was just a matter of time to make it official.

In the afternoon, she again met Dustin at Kara's place, this time taking Jordan. All of their relatives were now poring over Kara's little patch of five acres. They could see the remains of Kara's mobile home. It had been shredded into pieces and deposited into the ravine, near where Kara had been found. Sheet metal encircled what trees and fences remained. Mary picked through the trash, finding a broken unicorn statue that had been Kara's.

The Oklahoma Army National Guard joined the neighbors and volunteers in making a more systematic search of Bridge Creek. Captain Barry Guidry saw a little boy picking through the rubble.

"What are you doing?" he asked Jordan.

"I'm looking for my mom. You want to help?"

Guidry took Jordan by the hand and then put him on his shoulders. He gave Jordan his camouflage captain's hat to wear and another Guardsman gave him a canteen. The little soldier directed the search team around the property. He found his Matchbox cars, his baseball glove, missing pages from his baby book. But he didn't find his mother.

As their shift ended, the Guardsmen saluted Jordan as Guidry allowed him to dismiss the platoon. They gave him a combat patch.

On Thursday morning, the medical examiner made it official. Kara had been positively identified through the fingerprints. The relatives gathered at Mary's house and she sat Jordan on the couch and explained that his mother would not be coming back.

"How come?" he asked.

STORM WARNING

• • •

TIM MARSHALL, A CHICAGO native, became an engineer because of Ted Fujita. He grew up reading about the scientist's exploits in the local paper, and Fujita was his hero, second only to Chicago weatherman Harry Volkman. After Fujita published his groundbreaking research on the Super Outbreak, Marshall wrote him, requesting copies of all the scientist's hand-drawn maps. While other guys were posting pinups in their dorm rooms, Marshall hung Fujita's tornado track graphics. He attended Texas Tech for his graduate degree in engineering and worked at the wind research center before moving to a private firm in Dallas. But he retained his fascination with weather, particularly tornadoes.

And he still loved to chase. On May 3, he had called his wife at about noon and asked if she wanted to have dinner in Wichita Falls. "She knew what that meant." It meant a storm chase. The Marshalls eventually joined the parade of chasers on Interstate 44 as the giant tornado scoured its way into the metro area. Dinner that night was in Oklahoma City.

Marshall called Texas Tech the next day. He knew the wind research center would have an engineering team on site, and he wanted to be part of it. Texas Tech made him the director. The Texas Tech survey team was independent of the FEMA evaluation team, which included Chuck Doswell. But both surveys came to the same conclusion: numerous frame homes were poorly constructed. The minimal construction codes had not been followed. The lack of adequate shelter contributed mightily to the injuries and deaths.

"It's clear that the construction quality of homes over most of the United States is abysmally bad," Doswell said. "The home construction industry likes to pretend that all of their homes are well constructed. Well, what we found were numerous code violations, and the codes themselves are inadequate given the threat level associated with the climatology of tornados."

Marshall found one house where almost every nail in the plywood roof deck missed the underlying rafters. Gravity alone kept it atop the house—until the tornado. Rafters secured to the wall frame by nails met the minimal code requirements, but the metal hurricane clips that latch the rafters to the frame, used in coastal areas, are three times stronger than nail connections.

Some rural homes that did not have anchor bolts attaching the frame to the foundation were swept away nearly 300 feet. The 1,600-square-foot homes in Oklahoma City's Country Place addition were attached to slabs by tapered cut nails instead of anchor bolts. The nails penetrated the concrete by only half an inch. These homes left a debris trail of more than 3,000 feet into an open field. In no instance did Marshall and the crew find an anchor bolt that had failed.

They found numerous instances of garage door failure. Garage doors, especially those made of aluminum, easily collapsed and allowed the winds entry into the house, which hastened its destruction.

Marshall found a thirty-six-foot steel beam that the tornado twisted like a swizzle stick as it threw it nearly 1,000 feet. He found scores of cars rolled and tossed and barely recognizable. He found a long stretch of barbed-wire fence that the tornado rolled into a three-foot ball, a steel leg of a lawn chair that penetrated a solid five-inch-by-five-inch post, and a fifty-pound, six-foot-long section of sewer pipe that burst through the front door of a house and ended up in an interior hallway.

Marshall and his Texas Tech team interviewed 100 survivors. Each person had known the tornado was approaching because of warnings from television or calls from friends. Several had driven away from the area. "Although driving away from the tornado was a successful strategy, fleeing a tornado in lieu of seeking shelter in one's home is generally not recommended," Marshall wrote—unless the home is a mobile home.

Marshall discovered that a number of homes collapsed before the winds of the tornado reached their maximum. He concluded the homes disintegrated when the twister's outer winds reached F2 speed, 113 to 157 miles per hour, far below A9's top speed. He also returned after three months to review forty homes under construction. "The author found that, in general, the quality of the new construction was no better than the quality of construction of destroyed homes," he wrote.

The problem, said Marshall with a hint of irony, was that A9 was too powerful.

"We found violations here and violations there, and it pretty much got shoved under the rug. It was just too strong of a tornado. Now, if it was a weak tornado and you had houses falling down, then people would start to complain about this," Marshall said.

Most tornadoes were not 300-mile-per-hour F5 behemoths. A frame home properly strapped and bolted should withstand at least an F2 tornado, Marshall said. Those types of twisters are far more common in Tornado Alley.

He did discuss the issue of construction codes with Oklahoma officials before reconstruction started. "They kind of told me to go back to Texas."

And when it came to mobile homes, all bets were off. As the percentage of mobile homes increased in the United States, so did tornado deaths among mobile home residents. By 1999, half of all annual tornado deaths were mobile home residents.

Five years earlier, the Manufactured Housing Association of Oklahoma had lobbied the local National Weather Service to tone down its recommendation that mobile homes were not safe shelters during tornadoes. "They claimed that the NWS is 'unfairly' singling out mobile homes in warnings," said Harold Brooks at the National Severe Storms Lab. "Their position was that new manufacturing procedures and tie-down regulations have resulted in improved safety. While it is possible that the improvements in construction and installation may lead to a reduction in fatalities in mobile homes, such a reduction has not occurred yet."

The Norman meteorologists suspected the mobile home industry was conducting a national campaign to persuade the National Weather Service offices to change warnings and safety advisories. Kansas also was debating whether to require mobile home parks to construct underground shelters. The topic of shelters and trailers was becoming a political issue. The Weather Service kept its mobile home warning, urging people to abandon them immediately during a tornado threat.

Brooks, who has conducted extensive studies into tornado deaths, estimated that a mobile home resident is twenty times more likely to die than a resident of a site-built home during a tornado. Yet only Minnesota requires mobile home communities to have on-site shelters. Only two cities—Wichita, Kansas, and St. Joseph, Missouri—have similar shelter requirements for mobile home parks.

And there are many legitimate issues surrounding mandatory shelters. There are liability concerns. Owners of mobile home parks fear lawsuits. Even a seemingly simple question can be fraught with problems. Should pets be allowed? What if other residents are allergic?

Chad Erwin, Kara's neighbor, could not resist gigging the salesman who had sold him the double-wide mobile home. In making his pitch, the salesman said he had heard of only one double-wide home moving more than six inches because of the winds. Chad called him up after the twister. "My house just moved 200 feet. Can I have my money back?"

THE NATIONAL WEATHER SERVICE did a quick count of the numbers and sizes of the May 3 tornadoes. The statistics were staggering. It was the biggest outbreak of supercells and the biggest outbreak of violent tornadoes in the history of Oklahoma. A9 was the most powerful tornado ever to strike the metropolitan area. The Storm Prediction Center kept a database dating to 1950. In five decades, it had counted 40,522 tornadoes, but only 50 achieved the highest rating on the Fujita Scale, the F5. The Oklahoma City monster was the fifty-first. Its winds of more than 300 miles per hour made it the most powerful F5 ever recorded.

There were seventy-one documented tornadoes in Oklahoma on May 3. The instability rippled the atmosphere from Texas through South Dakota, producing ninety-six twisters throughout the Central Plains on that Monday and Tuesday.

There was at least one tornado on the ground somewhere for eleven hours. At one point, there were four twisters on the ground at once. There were eleven supercell formations, including one in Kansas.

The Norman office of the National Weather Service issued 116 tornado warnings. To put that in perspective, weather offices in thirteen states issued 150 warnings during the 1974 Super Outbreak.

In the metro area, 2,300 homes were destroyed, 7,300 damaged. There were 473 apartment units destroyed, 568 damaged. There were 164 businesses destroyed, including all 53 stores at the Tanger Outlet Center in Stroud, and 96 businesses damaged. Five churches and two schools were leveled.

The cost: $1.2 billion, the nation's first billion-dollar twister. Before May 3, the costliest tornado in Oklahoma City had been the twister that hit Tinker Air Force Base in 1948. Its tab was $10 million.

The only statistic missing from the record books was the number of fatalities. The tornado outbreak claimed 47 victims in Oklahoma and Kansas. A9 killed 38 people. More than 800 people in the metro area were injured. Four people died outside the Oklahoma City area; 5 were killed by a twister in Kansas, and 3 died indirectly from heart attacks or injuries related to seeking shelter.

Gary England was stunned by the low death toll. "I thought it was going to be in the thousands." To have 11,000 homes, apartments, and businesses damaged or destroyed and to have only 47 dead was almost as incredible as the tornado itself.

The previous record-setting 1942 tornado killed 31 people in Oklahoma City, destroyed 70 homes, and caused $500,000 in damages.

The unusually low death toll on May 3 didn't happen by accident.

Oklahomans heeded the high-pitched warnings from the television and Weather Service forecasters. Thousands either left the area or found shelter. They did, as Gary England warned them repeatedly, get below ground. For at least one afternoon in May, all the hard work, all the billions of dollars in warning systems, paid off. The Tinker crew's first forecast, Horace Byers's Thunderstorm Project, Ted Fujita's imaginative creativity, the National Severe Storm Lab's Doppler innovations, forecasts from the National Weather Service and the Storm Prediction Center, $4.5 billion spent for modernizing the weather agencies, England's own innovations, and Ken Crawford's OK-FIRST all built one upon the other.

And there was the matter of luck.

The national storm lab did some calculations. Had a tornado the massive size of A9 happened in a less twister-savvy state, had it been at night like B20, had its daylight movements not been carried live on the local television stations, had the warnings at all levels not been so forceful, clear, and repetitive, researchers estimated the death toll would have been 600 people.

The fatality study wasn't an intellectual exercise. It was intended to be a wake-up call for the nation. Tornadoes are dangerous and deadly, and they can happen anywhere. And tornado fatalities are likely to be higher outside Tornado Alley. In the twister-prone states, people are better educated about storms and shelters.

Another survey of those who were injured found that the vast majority knew of the warnings and the tornado. Like Kara Wiese, they just did not have a proper place for shelter. It also found that many of the injuries were life shattering. Thirty people, including nine children, had severe and permanent brain injuries.

THE DEATH TOLL MAY have been statistically low, but the misery was hard and deep. Twelve of the dead, the largest number of fatalities at one location, were from Bridge Creek. Terry Brown spent three nights on his office floor before he slept in his own bed. He became the de facto city manager of the unincorporated community. The Federal Emergency Management Administration's recovery efforts are designed to work with cities or counties, but many Bridge Creek roads were privately owned; large acreages made it difficult for home owners to get debris to a clean-up point. The government would not pay for clean-up operations unless the trash was brought to a publicly owned road.

A9 left an estimated 220 million cubic yards of debris.

It was hard enough in a fully functioning city. Moore assistant city manager Steve Eddy stood in the middle of one destroyed neighborhood feeling distraught. More than 600 homes in Moore had been obliterated, block after block of nothing but splinters and pink insulation. Eddy felt "an utter sense of despair."

Eddy would have plenty of offers of help. Several companies specialize in disaster clean-up operations, and they were calling him with offers. Instead, Moore went with local firms. Other metro area officials went through the U.S. Army Corps of Engineers, which issued contracts to private companies. Eddy thought it would take more than a year just to clear away the debris; the local company did it in eight months. It took Oklahoma City and the other suburbs months longer.

At Bridge Creek, the school became the command center. Kay Norman, the elementary school principal, became the supply chief. Janie Pruett became the cook, turning the high school kitchen into a restaurant for weary survivors. School board member Jim Seals became the communications chief, setting up a network of ham radio operators. The Highway Patrol, FEMA, and the Red Cross later set up operations at the school, but for a while, Janie was cooking for a couple hundred people three times a day.

With all the injured at hospitals and all the dead at the medical examiner's office, Kay and Janie grabbed mops and buckets full of bleach and water to clean the blood and body fluids from the school floors. They had just started when a dozen young men walked into the gymnasium. "Let us do that," one said, and grabbed a mop. They were from a religious relief organization.

Bridge Creek became inundated with volunteers from churches all over the country. Evangelical Christians, Southern Baptists, Methodists. Many were Mennonites who traveled from one natural disaster to another. Opposed to military service, the Mennonites had established a disaster relief organization during World War II as an alternative national service. The Mennonite Disaster Service, with 3,000 volunteers, became a little-known but critical player in disaster recovery operations.

Donated goods arrived by the truckload. Eventually, seventeen semi-trailers packed with goods sat in the high school parking lot. Kay and the teachers separated the supplies on folding tables placed in the gym. "That became the Bridge Creek Wal-Mart Supercenter," Kay said. The local Wal-Mart provided them with pushcarts.

In the middle school, they set up a recovery room—a place where people brought photographs or personal items found on their property miles from Bridge Creek. "There was a white wedding dress just covered in mud. It was so poignant to me," said Kay.

Superintendent Terry Brown visited a friend near the school who had lost everything. There was, said Brown, nothing left bigger than an ink pen. They used wire mesh to sift through the rubble, as though they were on an archaeological dig. As the dirt fell through the wire, a gold wedding band appeared. At least his friend had something. When Brown wasn't attending FEMA meetings in Oklahoma City, he was driving a school bus. The buses became the delivery trucks, hauling goods, supplies, and equipment around to the needy at Bridge Creek.

It seemed everyone had a horror story of survival. Counselors moved into the school to help the living cope with the tragedy. Just about everyone pitched in to help. Chad and a group of Mennonite students handraked all the pasturelands, gathering up trash and all that pink insulation. Dump trucks made hundreds of trips, clearing away the bigger sheets of metal and twisted cars, torn furniture, and battered refrigerators.

There would be no May graduation ceremony for the Bridge Creek senior class of 1999. For two weeks straight, the Bridge Creek school employees and volunteers worked nonstop, nearly twenty hours a day. At the end of two weeks, Janie took a break to join friends for a birthday luncheon in Oklahoma City. "I had never felt so weird at a restaurant in my life. I saw these cars going by and I thought 'All these people were leading a normal life. We're not leading a normal life.'" The adrenaline left her body. "I wanted to sit there and cry. Even my friends didn't know what we were all going through here. I wanted to say, 'You don't know what we're doing; you don't understand.'"

Five families, including the Erwins, lived in the elementary school classrooms throughout the summer, and each person would reach out or inside to find their way.

Brown told his pastor that he had given and given emotionally: "There's no me left." One of the outside counselors came into Brown's office and shut his door. "We're going to talk," the therapist said. Brown was initially reluctant. "I'm a macho man," he joked. But the two began talking once a week. The counselors stayed through the summer.

But not everyone could be helped. One survivor, who lived by hiding in his shower, dragged a kitchen chair down his driveway to his mailbox on the main road. He didn't trust the government and never sought any assistance. His wife had been in a nursing home for years, and his insurance was cancelled. He sat down, propped a shotgun to his mouth, and pushed the trigger with his thumb. A9 claimed one last Bridge Creek victim.

OVER THE DAYS AND weeks after the tornado, Charlie and Pam Cusack returned repeatedly to their splintered home to salvage what they could. Every trip ended the same way, with both of them crying, not only for themselves but for the massive devastation all around them. It wasn't until he emerged from behind that flimsy bathroom door that Charlie understood. All those tornado victims interviewed on television over the years, all saying the same thing: stuff can be replaced; I have my family.

His big-screen TV was smashed; all his electronic equipment was destroyed; the house was gone. But when he pulled his daughters from the rubble, he understood: "I have my family." It was, he said, a gut-wrenching experience.

The Cusacks built a new home, not in the same neighborhood. They moved to the east side of Moore and, when they built the new home, they made sure it included an underground storm shelter. "You didn't learn anything if you didn't learn that."

Charlie kept a souvenir from May 3 that displayed the enormous power of the twister. It was a snapshot given to him by his insurance agent. The picture showed a muddy terrycloth house slipper, the soft, rubber-soled kind. The heel of the house slipper was embedded one inch deep in the steel-belted radial tire of a Corvette.

That was one of the markers of an F5: "Incredible phenomena can occur."

Carolyn Stager lived with her parents for the next year. Recovery was a long, slow process, and she chronicled her experiences as a way to deal with her feelings. A pretty, petite, stylish woman, she sported a blonde wig, and her friends all remarked on her progress. They complimented her on the cute hats that she wore and her upbeat spirit. But it was hard. Over the next four years, she underwent repeated surgeries as doctors implanted balloons under what was left of her scalp. The balloons helped stretch the skin until it again could cover her head.

"The experience was horrific. It's like you are living one of those nightmares that you read or hear about happening to someone else. And then you realize it's happening to you. I think we all somehow feel immune, when we read or hear about those things. It's always happening to some distant person."

Her daughter had a difficult recovery as well. Years later, pieces of glass and metal still worked their way out of her body. But everyone lived. She found the things that were important to her. Her Bible always lay on her bed. It was found, still intact, blocks away in the yard of a woman who worked at the state capitol and knew Carolyn. Someone found her driver's license and mailed it to her.

She had three good pieces of jewelry that she kept in a blue sock in her dresser. Someone found that as well. Churches set up repositories where people brought photographs found on their property. Carolyn called it "Tuesday nights out with the walker" as she made repeated trips until she found some of her snapshots.

She endured difficult physical therapy to recover from the broken pelvis, and she recovered quickly. "It wasn't my style to be down and out." She sold the lot where her home of twenty years had been and moved to the opposite side of Oklahoma City. At middle age, she was starting life over again. But at least she had her family.

Gary England also was shaken by the twister. The veteran had warned Oklahomans of hundreds of tornadoes in his years as chief meteorologist. He often told his colleagues that one day, the big one would pay a visit, and when it did, it sapped him like nothing else ever had. He even considered retirement, briefly. The tornado was just too much to comprehend.

AT THE STORM PREDICTION Center, Roger Edwards and Rich Thompson pored over every little detail about May 3 that they could find. They searched through historical records about past supercell outbreaks. Nothing about the supercells on May 3 was textbook. Most major outbreaks were accompanied by extremes of instability, shear, moisture, and lift, but there was little lift available. Most thunderstorms form along the dryline or weather front, but the first few storms formed miles from the dryline and no front was nearby.

Reviewing the various Doppler scans at different altitudes, Rich found one clue: a thin, horizontal, convective roll, a little cloud plume that often forms from afternoon surface heating. Similar in appearance to a jet's contrail, the horizontal convective roll is a wind spiral that elongates like a stretched-out Slinky. Their best theory, at least for Storm A, is that the spiraling winds provided just the extra lift to burst through the atmospheric cap.

"That was where it got warmest; that was all you could say. It was warm and there was something, a little thing like the roll cloud. But in real time, that was not apparent," Rich said. "That affected Storm A, we think. We can't to this day tell you why the other supercells formed."

All these little things go into the forecaster's toolbox, said Roger. "One of the biggest lessons learned meteorologically was that these ingredients don't have to be classical or extreme to get a tornado outbreak. They have to be adequate and all present."

"Less can be more," said Rich. "We had been thinking more is better. Now when we look at it, subtle is better. We look for that kind of environment now. Weak cap, strongly unstable, weak lift, lots of moisture: that may be the recipe to give us the kind of storms that cause the most problems in terms of tornadoes. It may not be the easiest to figure out. But that may be how we get the isolated supercells that cause the most problems."

The SPC team also wrote an analysis of the forecasting difficulties of May 3:

> *Operational numerical forecast models valid during the outbreak gave inaccurate, inconsistent and/or ambiguous guidance to forecasters, most notably with varying convective precipitation forecasts and under forecast wind speeds in the middle and upper troposphere, which led forecasters (in the early convective outlooks) to expect a substantially reduced tornado threat compared to what was observed.*
>
> *That, combined with relatively weak forecast and observed low-level convergence along a dryline, contributed to much uncertainty regarding timing and location of convective initiation. Consequently, as the event approached, observational diagnosis and analysis became more important, and were critical in identification of the evolution of the outbreak. Ultimately, tornadic supercells developed earlier, were more numerous, and produced more significant tornadoes than anticipated.*

Their report was another nod toward the importance of the human factor in forecasting and a warning against an overreliance on computers. "This event serves to alert forecasters that excessive reliance on model guidance in such situations, at the expense of examining real-time observations, not only increases the potential for inaccuracy, but can be dangerous when the quality of watches and warnings for deadly events is affected," the report concluded.

None of the meteorologists, not at the Storm Prediction Center or the National Weather Service, not Gary England, predicted the strength of the storms, the sheer number of tornadoes, or their power.

For all of 1999, Oklahoma had 145 tornadoes, a new state record. It was a vicious tornado year in the most unexpected places. That August, a tornado tore through downtown Salt Lake City, making a rare appearance west of the Rockies and killing one person. Nationwide there were more than 1,225 twisters, the fourth highest number since NWS records began in 1950. In 1999, tornadoes killed 94 people nationwide.

Some people wonder if the climate was sending a cosmic signal. In 1953, killer tornado outbreaks raised concerns about the impact of nuclear tests in Nevada. After 1999, the number of outbreaks prompted questions about El Niño and the impact of climate change. Statistically, the outbreak was not out of the norm. It was, said some scientists, just a freak of nature.

THE DAY AFTER THE tornado, the meteorologists in Norman, those from the National Weather Service, the storm lab, and the Storm Prediction Center, wanted to make one message crystal clear: highway underpasses are the worst places to be during a tornado.

Three people had died at underpasses. Drivers stopping along the interstate highways to seek shelter had blocked traffic. Other people caught in the traffic jam were trapped when the tornado arrived. There was even one couple who left their home and drove to an underpass thinking it would be safer.

Texas Tech had been talking about shelters for years. It had worked with FEMA to create a booklet published in 1998 explaining the parameters of safe shelter. For days after the Oklahoma outbreak, the issue of shelters was the main topic.

President Clinton made the kind of endorsement the Texas Tech engineers had been waiting decades to hear. "For goodness sake, build a safe room in your house when you rebuild. It will be the cheapest $2,000 you ever spent," Clinton told Oklahomans. More than 6,000 metro residents, encouraged by a FEMA shelter rebate, did just that.

Texas Tech's Ernst Kiesling formed the National Storm Shelter Association after the Oklahoma outbreak. The idea was to set some national standards for shelter construction and to serve as a seal of approval for consumers.

Doswell noted that the entire warning system on May 3 worked incredibly well. Some people had a thirty-minute warning time. Yet there were fatalities. The number of deaths was probably the lowest possible number given the power of A9.

Doswell argued that extending the tornado warning time may not be the answer to saving more lives. The tornadoes causing the most deaths are those like A9—powerful, long-tracked, massive tornadoes that also allow forecasters to provide long warning times.

Tornado fatalities in the United States usually fall into three categories: those in mobile homes, those in cars, and those in homes without shelters.

"I think we're close to the minimum number of fatalities associated with tornadoes these days," Doswell said. What people lack are adequate shelters. Better construction of homes would be a major improvement. But he believes it's virtually impossible to change construction codes if home builders are opposed.

In-home or in-ground shelters were popular after May 3, but researchers wonder if the trend will continue. There were some calls for large public shelters, similar to the old fallout shelters of the Cold War, but local officials feared more people would be injured driving to a public shelter than if they just took cover at home.

A9 truly rattled the people of Tornado Alley like no other twister— or at least no other that they remembered.

Survivors' fears were deeper than just having a shelter. The city of Moore stopped its weekly testing of its tornado sirens for nearly a year to keep from unnerving its residents. Oklahoma City residents passed a bond issue that financed state-of-the-art tornado sirens to replace its Cold War–vintage alarms and built a large emergency command center that could withstand F3 winds.

At Bridge Creek, Terry Brown received additional funds to build more elementary school classrooms and made each one a reinforced shelter that would serve not only the students but the community.

But the first time Brown attempted to keep students after school when there were high-wind warnings, parents ran to the school to pick up their children. They wanted their family with them, not trusting anyone else with their child's safety. "I can understand that," said Brown. "But the kids would have been safer here."

Bridge Creek itself was changing. The county built a new highway to the west side of Bridge Creek that made it far easier to find the community. A portion of the older housing additions voted to incorporate so Bridge Creek would have a mayor. And the empty pastureland north of the school filled with large three- and four-bedroom brick homes, and most of them had shelters.

The numbers of mobile homes south of the school are much sparser. Chad Erwin bought a new double-wide, but the lot next door, Kara's place, remained vacant. Chad also built an in-ground tornado shelter, and inside he placed his own talismans: a piece of the green carpet and tree limb that helped save his family.

ON A HOT AUGUST day in 2001, Mary felt a lump on Jordan's lower left leg. She took him to the hospital where she worked to have it checked out. It was a tumor, the doctors told her. It was cancerous. They needed to amputate below the knee.

The little boy who had survived the most powerful tornado ever recorded had to battle again. He underwent chemotherapy, and all his snow-white hair fell out. It grew back dark and coarse, just like his mood.

Mary's own mother had died the previous February, and she carried the grief for her mother, her daughter, and her grandson. "My mom was just such a worrier. I was almost glad she had died before and didn't have to see him with cancer. That worked with her. I tried to do the same thing with Kara. I tried to be glad that Kara didn't have to see him go through that. But it didn't work."

She missed her daughter terribly, and the burden of caring for not just a little boy, but a very sick little boy, was overwhelming. There had not been time for Jordan to adjust to the death of his mother before he was diagnosed with cancer and lost his leg.

His own grief turned to anger, his sweet and smart temperament flecked with rages he found difficult to control. "It's like everything has been an emotional overload for him," Dustin said.

Perhaps, suggested Mary, it is partly her fault, so distraught was she after Kara's death. "He came to live with me, and the truth of the matter is that was probably the worst thing for him right at that point. I was in no shape to start raising a little boy. No matter how much I wanted to do it. I was maybe the place where he felt safest but also the place where he got away with murder."

Perhaps that's the lingering power of a tornado: the damage continues long after it's gone. School psychologists who studied the effects of tornadoes and hurricanes on children pointed to the randomness of a tornado; the unpredictability undermined children's sense of safety and security. The experience carved itself into the child's brain, leaving vivid memories. Children often retold stories of grotesque images. A profound sense of loss overwhelmed them emotionally, creating unreasonable fears and behavior difficulties. And the effect could be long lasting. It had been for Jordan. Five years after the twister, he was still talking about that dead horse he saw being killed. And his mother—the memory of her face was beginning to fade.

"We had no respect for tornadoes, no respect at all for what they could do," said Mary. "We have respect now."

13

SEEING THE WINDS

R ARELY DOES AN ERA END SO neatly, so abruptly.
In March 1998, Tinker Air Force Base and the National Weather Service hosted a fiftieth anniversary gala to mark the achievements of Ernest Fawbush and Robert Miller. Fifty years earlier, they had peered into their own surplus radar scope and sketched the dangers on their acetate charts and swallowed hard and made the first tornado forecast.

At Tinker, they dedicated a red granite monument:

First Tornado Forecast

MARCH 25, 1948

This memorial is dedicated to the first operational tornado forecast issued on March 25, 1948, by Major Ernest J. Fawbush and Captain Robert C. Miller at Tinker Air Force Base, Oklahoma.

Issued several hours before a tornado struck Tinker Air Force Base, this first forecast proved severe weather could be anticipated with a reasonable degree of accuracy. This focused national attention on forecasting tornadoes and warning the public of their potential danger.

Seeing the Winds

Severe weather pioneers, Major Fawbush and Captain Miller developed tornado forecasting techniques still in use today. The 1948 tornado forecast was the forerunner of today's national severe weather forecasting and research program that protects lives and serves the American people.

Dedicated March 25, 1998

NOAA held an open house to mark the occasion, swinging open the doors to the National Weather Service office, the National Severe Storms Laboratory, and the Storm Prediction Center. The U.S. Postal Service issued a special stamp. Oklahoma schoolchildren competed in poster contests on tornado warnings and safety. There was a symposium on storms at the University of Oklahoma and a fly-by of Air Force jets.

Fawbush had been dead for more than a decade, and Miller suffered from Parkinson's disease. Both were represented by family members.

NOAA had reason to celebrate and show off its new equipment. The $4.5 billion modernization project, four times the initial cost estimate, was finally completed for the National Weather Service. The last of the 116 Doppler radars became operational in 1998. NWS also shuttered nearly half of its local weather offices, so the budget cutters were happy as well.

But on this spring day, it was all about celebration and partnerships and perhaps a little silent apology. It was also a time for reflection.

It really had been an amazing fifty years.

The year 1998 served as the bookend to 1953—the last time a single tornado killed 100 or more people, the year of Fujita's arrival in the United States. In 1998, the deaths all came quickly—a generation of pioneers was gone.

First was Horace Byers, who directed the Tornado Project and helped bring Fujita to the United States. After Byers, there was Miller, who had once taken such glee in one well-crafted weather chart and in besting the civilians. And, finally, in November, Fujita.

Fujita had retired from teaching in 1990 at age seventy but still engaged in research. His main passion was completing his book: *Mem-*

oirs of an Effort to Unlock the Mystery of Severe Storms, which he self-published in 1992.

Ever the researcher-engineer-cartographer, Fujita turned to himself. He crafted one chart depicting the price of rice in yen, the price of rice in dollars, the yen to the dollar, the Dow Jones Industrials, and the Consumer Price Index (CPI) from his birth in 1920 until his retirement in 1990. "Although this logarithmic diagram does not represent the variation on natural coordinates, logarithmic values are excellent in depicting percent changes," he wrote.

He charted the annual amount of his grants versus the CPI. Fujita received more than $12 million in grants during his years as a researcher.

In 1983, he had purchased a blood pressure gauge and used it regularly with his own little experiments, such as testing his pressure before and after an evening beer, mowing the lawn, or drinking hot tea. Blood, like the wind, had its own currents and force.

Brian Smith, one of Fujita's grad students, recalled that his mentor became more sociable after retirement. "Instead of him being too busy to talk to me on the telephone because he was so involved in his research, it became increasingly hard for me to get him off the telephone," Smith said. "I believe he felt the pressure to produce accomplishments in meteorological research was over. He could relax more."

Fujita readily accepted speaking engagements, including Chicago WGN-TV's annual tornado presentation at Fermi Laboratory. His accent still heavy, Fujita nonetheless captivated the crowd of more than 1,000 people, most of them ordinary folks who had an interest in storms. Afterward, children surrounded Fujita to have him sign their program "Ted Fujita, Mr. Tornado."

He attended the sessions to encourage youngsters to become meteorologists. Who knew which child might become the one to break the tornado's secrets?

Greg Forbes, another of Fujita's grad students and the Weather Channel's storm expert, recalled that Fujita never discussed his personal life with students. Fujita's son, Kazuya, said his father's personal life was his professional life. His research always occupied him.

Still, Forbes and others were surprised when Fujita, whom Forbes always referred to as Dr. Fujita, made a presentation for his memoirs. The students had never heard of Fujita's damage survey in Hiroshima or Nagasaki. They were unaware of his narrow wartime escapes from the fire bombing or the bomb itself.

After retirement, Fujita spent much time in his backyard, photographing and studying his "silent friends," praying mantises. He was a naturalist at heart.

"I recall that I was able to walk on a series of stepping stones marked 'good luck' ever since I was fished out of postwar Japan to the University of Chicago by Professor Horace R. Byers, my fatherly mentor professor," Fujita wrote.

Life, as Fujita recalled his father explaining, is ever changing.

He was a man who found order amid the chaos, both natural and man-made; a gumshoe worthy of the most classic pulp fiction, a diviner of mysteries, and a saver of lives—though people might not even know they had been saved or that they even needed saving—from tornadoes and their downburst kin.

It was Fujita who gave forecasters so many tools, so much knowledge: the mesocyclone, the supercell cycling, the multiple vortices, the relationship of the hook echo, the moving satellite pictures, the correlation of cloud tops to tornado production, the downburst.

"Ted always said that he wanted to leave behind a legacy," said Roger Wakimoto, Fujita's former graduate student. "He attained this goal. Ted's interests were diverse, and he would have succeeded in almost any scientific field with his immense curiosity, intuition, uncanny ability to conceptualize, and his tireless work habits. We were lucky Fujita chose meteorology."

The American Meteorological Society in the fall of 1998 began planning an unprecedented event to honor Fujita for his lifetime achievements. Usually such a major symposium honoring one scientist is held posthumously. But Fujita, whom *Weatherwise* magazine called "probably the greatest meteorological detective who ever lived," would be different. The event was planned for January 2000.

"I was thrilled to call Ted and inform him of the good news. He was deeply touched and humbled by the honor but also told me that he would not be able to attend. I understood the implicit meaning behind his reply, but, nevertheless, was deeply shocked and saddened when he passed away the following month," recalled Wakimoto. Diabetes had forced him to his bed in his final days.

The University of Chicago had closed Fujita's Wind Research Center. There was no successor to him. The university dismantled Fujita's tornado simulator, though Texas Tech's Jim McDonald sought desperately to obtain it for historical display at the Wind Science and Engineering Research Center. There was no one to update his historic database of tornadoes. There was no one who could command a Cessna on a moment's notice to spiral around the plains; there was no one to examine damage paths or starbursts. There was no one to see the winds.

There would be a Buddhist funeral for Fujita, a small affair with fifty to sixty of his friends and colleagues. And later he was buried in Japan, in Nakasone, in the family cemetery.

The golden era for tornado research ended in 1998. Like a long-tracked twister, it disappeared abruptly but left behind a trail of knowledge for others to follow.

The AMS held its Fujita symposium in California in 2000, and there were many presentations on Fujita's work. His son and second wife, Susie, were there for the tributes. But the last day, the meteorologists devoted to the 1999 Oklahoma tornadoes. No doubt Fujita would have been intrigued by those powerful twisters, so it seemed a fitting addendum to the symposium.

Fujita's observational methods slowly gave way to more sophisticated tools. Storm research was evolving and maturing, thanks to Doppler radar and great leaps in computer technology. Supercomputers, capable of calculating billions of data per second, assisted in creating weather models for forecasters and models for storm analysis.

Fujita had warmed to the Doppler but the numerical modeling computer was a different matter. He drafted his own models as he always had, by hand. How could the computer see the winds? "The computer," he said, "doesn't understand."

14

A Tornado's Grip

THE MEATWAGON MET A HORRIBLE END. Roger was at the wheel, Rich in the navigator's seat. They were taking an exit ramp on their way to chase a storm when the Meatwagon hydroplaned and straddled a concrete median, which cleaned out the entire underbelly, including the axle. It was a sad end to a trusty steed, a reliable storm intercept partner, and a willing fishing companion. Roger promptly bought a used Ford Crown Victoria, the favorite of police officers and taxi drivers because of its durability, so he and Rich could keep searching for violent storms.

Joining the Meatwagon on the junk heap was NOAA's Cray 90 supercomputer, the Maryland-based weather brain that produced the numerical modeling programs. That September, it caught fire, and the first fireman on the scene hosed it down with an extinguisher, frying all its circuits.

In its dry language to its forecasters, NOAA noted, "The Cray 90 is inoperative." It was replaced with an IBM supercomputer that was five times faster than the four-year-old Cray model, which spoke volumes about the growth of supercomputer technology.

Once national weather forecasters depended on ground observers to send them the daily temperatures, air pressure, and humidity. Now space-based satellites, marine buoys, worldwide balloon soundings, Doppler radars, and wind profilers spew out 130 million weather factoids daily. NOAA's computer can make 450 billion calculations per second, deciphering weather data for 6 billion cubic miles of atmosphere around the world and crafting forecasts from the surface to 30 miles high.

But what it still cannot do is tell forecasters where a tornado will be or why. The supercomputer is limited by meteorology's lack of knowledge. And it lacks a forecaster's intuition.

In Oklahoma, the fun and collegiality within the chase community was fraying amid a backlash. Some people thought there was too much glee on the part of some chasers at seeing such a huge, murderous twister. Roger weighed in on the storm chase etiquette, criticizing any chaser who would applaud and cheer at such a horrible sight, but also defending the chasers who have respect and awe for nature's creation.

"I refuse to feel a shred of remorse or guilt for my fascination with severe storms. I have no reason to whatsoever, because they are powerful and fascinating, they're the reason I do what I do," Roger said. "That fascination, in turn, leads people like Chuck Doswell and Rich and I into severe storms to try to mitigate the damage and loss of life that they cause. For storm spotters, the fascination leads to the motivation to go out and spot, and, in turn, to call in reports to the Weather Service, which, in turn, may save lives."

He offers no defense for that small group of chasers who clap and yell at the sight of a tornado and understand why people would be upset with chasers in general.

"Sometimes there are just flat-out too many people out there. The roads get clogged; emergency vehicles have trouble getting through. And so I understand why some folks get pissed off at chasers and spotters, because they see us—and by 'us,' I'm putting on the coat of a chaser and not of a meteorologist—they see us as outside interlopers who are celebrating their problem. And the problem is the storm."

● ● ●

THE 1999 OUTBREAK ONCE again focused the federal government's attention on storms. The Oklahoma Mesonet and OK-FIRST projects received national recognition as Harvard's Kennedy School of Government in 2001 named it one of the five most innovative government programs. A 2002 presentation by the Oklahoma Climatological Survey drew representatives from twenty-five states plus Brazil, Finland, Canada, and Nepal.

Ken Crawford was named by NOAA to head a research project that would expand the Oklahoma Mesonet concept nationwide. The needs of each state would require some tweaking, but the idea is the same: putting real-time data into the hands of local emergency managers who could more quickly make the calls—to activate the storm sirens or evacuate an area—that so often save lives.

President Bill Clinton toured Del City after the May 3 twister and pronounced it the most devastated scene he had ever witnessed. He also erroneously attributed creation of the Doppler to the University of Oklahoma instead of the government's own National Severe Storms Laboratory. But perhaps he made up for that error by committing the government to construction of a new National Weather Center on the university campus.

The $67 million, seven-story facility was supported half by the federal government and half by the state of Oklahoma and the University of Oklahoma. The National Weather Center, which opened in the spring of 2007, houses the Storm Prediction Center as well as the national storm and radar labs. It provided a new home for the local National Weather Service office. And almost all of the university's storm research operations moved into the joint housing. The seventh floor includes an enclosed observatory from which scientists can survey miles of flat prairie. It is the perfect place to watch a supercell develop.

Roger Edwards drove past the construction site every day on his way to work, and at first it made him wince. "There's a lot of glass," he observed. But there also was a huge, reinforced storm shelter on the first floor.

The National Weather Center project also was a sign that Oklahoma's political leaders, perhaps for the first time, saw that there was money to be made in all those storms. They would make lemonade from nature's lemons. The National Weather Center would bring more than 650 employees under one roof and inject an estimated $45 million annually into the state's economy.

There was money to be made in the weather. And there was a hope that the National Weather Center would attract even more private sector weather business ventures into the state. Norman could officially become the weather weenie capital.

But the private sector's weather-related companies can be a troublesome lot for the Weather Service. Rick Santorum, a U.S. senator from Pennsylvania, sought to bar the Weather Service from providing free weather data to the public, with an exception, of course, for watches and warnings.

Santorum acted at the behest of private sector Pennsylvania companies that provide forecast services. Unlike the 1980s, when many in Congress seemed ready to cleave apart the Weather Service, Santorum's bill received no support.

The Pennsylvania Republican did receive a rash of criticism from the public and the press. Perhaps the public still thought the Weather Service was one government agency from which they received their money's worth—all for the price of a hamburger and fries. Charlie Cusack certainly thought so.

A FORMER RADIO DJ, Charlie kept the radio blaring all day at work. His new two-story brick home, complete with storm shelter, was a short two-mile drive away from the pet food store he owned in Oklahoma City. It was May 8, 2003, four years since the May 3 tornado destroyed his old home.

Suddenly the music stopped and the announcer reported that a tornado warning was in effect for Moore and south Oklahoma City. Charlie called home, where his fourteen-year-old daughter answered the phone.

"Honey, are you listening to the radio or watching TV? Because there's a tornado warning."

"No, I'm not."

"Let me talk to your mom."

"Well, she's not here." The dad in Charlie kicked in. There was no dillydallying this time. No big-screen TV viewing. "Oh, my Lord. I'll be there in a minute."

Charlie arrived just as his wife, Pam, drove up with another one of his brothers. The Moore sirens were bellowing. The Cusacks dashed toward their new shelter.

"So we're going down in the shelter and my brother said, 'I think I'm going to stand out here and finish smoking this cigarette.' And I looked at him, and I looked behind the backyard, and there was that tornado."

A Weather Service spotter first recorded the twister on the west side of I-35 and followed it as it moved down Twelfth Street in Moore, Charlie's old neighborhood, and crossed the interstate near Shields Boulevard where Tram Thu Bui had lost her life. It looped around south Oklahoma City, turning north again and moving across the southeast side of Tinker Air Force Base. Its winds caught part of the General Motors plant, damaging one building, and it moved on to open land.

The F4 twister missed Charlie's new house by a mile, but the one mile seemed too close for comfort. "The only thing I could think of was 'Here we go again.' "

It was the last of three tornadoes from a supercell that had been well predicted in advance. The National Weather Service's Norman office issued live briefings through NOAA Weather Radio nearly three hours before it touched down.

Gary England also was live on air. By 2003, Gary had an even more powerful Doppler radar that allowed News 9 to track the storm. Gary called his new Doppler MOAR—Mother of All Radar. He needed it.

For several miles, the 2003 tornado tracked parallel to and at times overlapped the path of the May 3 twister. The odds must be astronomical, but so was that entire week. From May 4 through May 11, 2003, an incredible 386 tornadoes were reported in twenty-one states and caused forty-one deaths. During that week, more than two dozen weather offices issued 5,124 warnings, including 1,115 tornado warnings.

Never before had so many tornadoes occurred in a one-week period.

STORM WARNING

• • •

THE MAY 3 OKLAHOMA outbreak and the destruction caused to so many residential neighborhoods also provided the initiative for experts to revisit the Fujita Scale. The Fujita Scale stood for more than thirty years as an imperfect measurement of the tornado. Led by Texas Tech's Wind Science and Engineering Research Center, a committee of forty engineers, storm researchers, and tornado experts finally refined the scale to better reflect the true powers of the winds and the damage they can cause to homes and lives.

They called it the Enhanced Fujita Scale, the EF Scale. To reduce the subjectivity, they based the measurements on twenty-eight indicators, taking into account, for instance, the damage caused to a poorly built home versus a well-built home. The committee report noted that a wind of 260 miles per hour—an F4/F5—was not required to destroy a well-constructed home and blow away debris. Homes could be, and were being, destroyed with a much less powerful wind. It also based wind speeds on three-second gusts, as opposed to Fujita's original quarter-mile measurement.

Starting in 2007, the National Weather Service will use the EF Scale, copying the same numbering system as the original but with different wind speeds.

Fujita Scale (mph)		Enhanced Fujita Scale (mph)	
F0 :	40–72	EF0 :	65–85
F1 :	73–112	EF1 :	86–110
F2 :	113–157	EF2 :	111–135
F3 :	158–207	EF3 :	136–165
F4 :	208–260	EF4 :	166–200
F5 :	261–318	EF5 :	More than 200

The new F5 will have no upper range, indicating that after 200 miles per hour, it becomes nearly impossible to estimate wind speed based on what little debris remains and that even well-built homes will be swept away. The Enhanced Fujita Scale, with its lower wind speeds, may even prompt people to wonder why their homes are not better constructed. Perhaps the next time, and there will be a next time, the numbing topic of construction codes will be the message.

The Federal Emergency Management Agency's Building Assessment Report on the May 3 storms provided a call for action with updated construction codes and enforcement. "Mitigating future losses, however, will not be accomplished by simply reading this report; mitigation is achieved when a community actively seeks and applies methods and approaches that lessen the degree of damage, injuries, and loss of life that may be sustained from future tornadoes."

Its conclusion: "Failures observed resulted from windborne debris and high winds that often produced forces on buildings not designed to withstand such forces. Failures, in some cases, also were observed that were due to improper construction techniques and poor selection of construction materials. Damage, in some situations, could have been reduced or avoided if newer building codes and engineering standards that provided better guidance for high wind events had been adopted, followed and enforced."

ON THE AFTERNOON OF May 3, Erik Rasmussen lost contact with the Mobile Mesonet crew. He couldn't get through on their cell phones or even contact the Norman Weather Service office. All the circuits were overloaded. Only later would the volunteers regale him with their tales: there were so many tornadoes they couldn't count them all. The VORTEX-99 crew, which mainly stuck with Storm B, recorded thousands of pieces of information on twelve tornadoes in just a few hours, far more than the original VORTEX armada did during its two years.

Such is the fickleness of tornado research.

"It is hard to gather the data you need on a rare phenomenon. In VORTEX, we had all these platforms, we had a real good plan on how to operate them, but when you get the data in your hands, you see that certain critical pieces are missing every time. If someone stopped their car to take pictures or a piece of equipment failed, when that happens on a rare phenomenon you don't have what you need," said Rasmussen.

Already Rasmussen and research scientists are working on VORTEX2, which, if it receives funding, would begin in 2008. It again would gather the premier scientists and all the latest gadgets, the newest and most sensitive radars, seven mobile Dopplers, a dozen mobile Mesonets, a couple of radar-carrying airplanes. VORTEX2 even plans to use computer-controlled aircraft, similar to the military Predator drones, but with six-foot wingspans that can be used to fly at low altitudes above chase cars. VORTEX2 also will study the effects of climate change on tornadoes to determine if there is a cause-and-effect relationship.

The original VORTEX cast a wide net over the entire supercell. VORTEX2 will narrow its studies to the storm's back side plus what is happening closer to the ground, ten meters and above, especially regarding temperature and humidity. Surprisingly, researchers have little information on temperature and humidity around the storm's back side. They have millions of observations about the wind for every one observation on temperature and humidity.

VORTEX-99 data from May 3 appeared to confirm suspicions that the tornadic action starts within the rear flank downdraft and near the ground, the first kilometer above the surface. When the downdraft was too cold, either no tornadoes or weak tornadoes formed. When the downdraft temperature was just right, just cool enough, it performed its powerful vortex dance with the spinning warmer ground winds lofted upward and a vicious funnel appeared.

"Before the tornado forms, it appears that the downdraft completely engulfs this developing vortex, which is bigger than the actual tornado. If the air in the downdraft is cold, it's much harder for it to be lofted because it's dense. If it's warmer, it can be lofted more easily, which means you can stretch that column of rotating air vertically and contract it horizontally into a tornado," explained Rasmussen.

Fujita had been the first to wonder about this downdraft. He called it a twisting downdraft. He was just thinking out loud. Possibly because of his frustration with the publication review process, he never officially wrote of it. He left future researchers with only a few tantalizing insights.

"That's really, really close to what we're seeing now. I don't think he had an explanation for it, but he did seem to ascribe tornado formation to this twisting downdraft which was then subsequently ignored, like many of Ted's ideas, for a couple of decades," Rasmussen said.

Mr. Tornado may have been close to solving the greatest of all the storm's mysteries.

IN HOUSTON, THERE LIVES a pleasant woman with a grandmotherly voice who is a fatalist about life. That is what the 1947 Oklahoma tornado did to Ramona Kolander when she was eighteen.

"I think when your time comes, you go. I have seen people survive very strange things and other people who have died immediately." Ramona Kolander is now Ramona Esphahanian, the mother of two and widowed since 1996.

She's always been a little dismayed that the 1947 twister is called the Woodward tornado. It affected so many people that day. There was the tiny Texas town of Glazier, where only the jailhouse was left standing. There was Higgins, where a friend of her brother was among the fifty-one people killed when a movie theater collapsed. And there was Shattuck, where her mother and brother Doug were the only casualties.

"There were a lot of people who died, and it didn't seem like they were acknowledged," Ramona said.

But she was talking about fatalism, the inability of people to affect their own destinies. Her little sister, LaNita, survived the twister outside Shattuck. Shards of glass and splinters worked their way out of LaNita's body for years afterward. Yet she died of breast cancer at age forty-three. Their seriously injured father, who had a broken hip and was in a cast up to his chest for months, lived to be ninety-seven. And then there's Ramona's own life.

"It set me on a path I would have never taken," she said of the '47 tornado. She had her own vision of her future. Had it not been for that tornado, she imagines she would have married someone she met at school, she would have remained in Shattuck, she would have raised her kids there, and now she would be retired in Shattuck.

Instead, she ended up living in several places in Texas and married an Iranian man, Nasser Esphahanian. She was working for the Soil Conservation Service in Texas, and he was a newly graduated oilman. They met in East Texas.

"We ended up living in Iran for about seven years and, gosh, we made trips around the world." They left Iran before the fall of the shah. But she saw a world outside Shattuck she never thought possible. "That never would have happened without that storm. I think things pretty much happen for a reason, but I can't always find what that reason is."

Ramona Kolander understands Jordan Wiese's guilt. The guilt can be almost as powerful as the tornado itself, tearing at the conscience. Jordan was holding his mother, Kara; Ramona was holding her brother Doug, and both were ripped from their grasps. In the end, there was no one to blame but coils of wind that lacked meaning or intent and the cosmic odds of 12 million to 1.

Ramona knows there was nothing she could do, but she, like Jordan, had been inside the bear's cage, and part of her never left. For more than fifty years, no matter where her life has taken her, she has had one unshakable thought: "I should have held that baby harder."

ACKNOWLEDGMENTS

I AM DEEPLY INDEBTED TO THOSE who were generous with their time and knowledge. Meteorology is a complex science and an art. The forecasters and researchers with the National Weather Service, Storm Prediction Center, National Severe Storms Laboratory, the University of Oklahoma, and KWTV Channel 9 patiently and passionately explained their work not just once but several times. Special thanks go to Roger Edwards, a forecaster with the Storm Prediction Center, who read the manuscript for scientific accuracy but proved he knows as much about grammar and geography as he does meteorology. Any errors of fact, however, are mine. Keli Tarp handled pestering questions with aplomb. Kazuya Fujita was gracious with his father's work.

But those who contributed to the heart of this book were the people caught within the swirl. Among them: Charlie Cusack, Carolyn Stager, and Jordan Wiese. The entire Wiese family was extremely helpful. Despite the pain I know it caused, they were eager to talk about their daughter, sister, and mother, Kara, who was such a vibrant light in their lives. And, hopefully, that light will help Jordan find his way.

As I hope this book makes clear, each horrific tornado event has been a learning experience not only for meteorologists who study the weather but for engineers who study the winds. The May 3, 1999, tornadoes—as well as Hurricane Katrina—underscored the need for stronger building codes and better shelters. Anyone interested in storm shelter standards or protecting homes from extreme winds should visit www.wind.ttu .edu for Texas Tech University's Wind Science and Engineering Research Center, www.nssa.cc for the National Storm Shelter Association, or www.fema.gov for the Federal Emergency Management Agency.

I also want to thank my friends who read so many versions of this

ACKNOWLEDGMENTS

book: Mary Reed, Paula Rubin, Chai Feldblum, my book club (Maria Cecilia, Marjorie, Margaret, Edie, and Amy), Debra Blake, Ellen Cornett, Roxi Slemp, Jim Myers, Barbara Moulton, and, last but not least, Ben Roth. Thanks to my coworkers and the concept of LWOP. The world's best steno and buddy, Ellen Eckert, who not only transcribed all taped interviews but offered valuable suggestions about the manuscript.

This book would not have been attempted without encouragement from Tony Dresden and inspiration from the late Christopher Marquis. The editors at Touchstone, especially Brett Valley, made this first endeavor a pleasant experience. Thanks also to Doris Cooper. Bob Mecoy of Creative Book Services is a great agent but a better friend. He kept the vision when I became lost. I miss the holiday dinners with Bob, his sister, Laura, and his mother, Alice. But Bob showed he still can turn a turkey into a decent meal. And, last, thanks to my family, whose love and support sustain me.

NOTES

G OVERNMENT WEATHER AGENCIES HAVE CHANGED NAMES numerous times in the past century. The U.S. Weather Bureau changed its name to the National Weather Service in 1967. The Storm Prediction Center (SPC) was born as the Severe Local Storms Unit (SELS), became the National Severe Storms Forecast Center, and changed to the SPC in the mid-1990s. To avoid confusion, the center is called the SPC throughout the book.

INTRODUCTION

x The breakthrough happened: James Gleick, *Chaos: Making a New Science* (New York: Penguin Books, 1988), pp. 16–18; Edward N. Lorenz, *The Essence of Chaos* (Seattle: University of Washington Press, 1995), pp. 134–136.

CHAPTER 1: NATURE'S ATOM BOMB

1 April 9, 1947, sometimes: Author interview with Ramona Kolander Esphahanian, March 2005; Donna Dreyer, *The 1947 Woodward Oklahoma Tornado* (Woodward, OK: Northwest Oklahoma Genealogy Society, 2000), available at www.usgennet.org/usa/ok/county/woodward/intro.html.

3 The state was a study: Jane Fitzgerald et al., *Partners in Flight: Bird Conservation Plan for the Osage Plains* (Washington, D.C.: Bureau of Land Management, Oct. 2000), p. 1.

3 About 140 million years ago: U.S. Geological Survey, "Interior Plains Province," in *Geological Provinces of the United States*, www2.nature.nps .gov/geology/usgsnps/province/intplain.html.

3 The vast middle of Oklahoma: Fitzgerald et al., *Partners in Flight*.

3 As America neared: Oklahoma Historical Society, *The Encyclopedia of*

Oklahoma History and Culture, 2 vols. (Oklahoma City: Oklahoma Historical Society, Nov. 2006), www.okhistory.org/enc/frame12.htm.

4 It was named for George Shattuck: Shattuck Chamber of Commerce, www.shattuckchamber.com.

4 Broom corn: ePodunk.com, Shattuck Community Profile.

4 The 1930s were the hottest: *Weather Time Line for Oklahoma, 1900 to 2000* (Norman, OK: Oklahoma Climatological Survey), p. 3. The 1950s would be hotter and drier but would lack the calamitous impact of the 1930s.

5 Black Sunday: *Weather Time Line for Oklahoma*, p. 3.

5 The thunder clapped: Interview with Kolander Esphahanian; Dreyer, *The 1947 Woodward Oklahoma Tornado*; impressions of storm also from *The Daily Oklahoman* and *The Dallas Morning News*, April 11–13, 1947.

5 In Woodward, war veteran Jim Feese: Dreyer, *The 1947 Woodward Oklahoma Tornado*.

6 The U.S. Weather Bureau had banned the word *tornado*: Roger Edwards, Storm Prediction Center, "The Online Tornado FAQs," www.spc.noaa.gov/faq/tornado/.

6 It started at White Deer: Dreyer, *The 1947 Woodward Oklahoma Tornado*; *The Dallas Morning News* and *The Daily Oklahoman*, Apr. 11, 1947.

6 blew nineteen boxcars: *The Dallas Morning News*, Apr. 10, 1947, p. 1.

6 A newspaper reported: *Shattuck Northwest Oklahoman*, Apr. 11, 1947, p. 1.

7 A nationwide telephone strike: "Hope of Phone Peace Dims as Southwestern Bell Quits Negotiations," *The Daily Oklahoman*, Apr. 10, 1947, p. 1.

7 A telephone operator in Shattuck: Ray Parr, "Woodward Toll Is 83 Dead," *The Daily Oklahoman*, Apr. 11, 1947, p. 1.

7 At 8:42 p.m.: Donald W. Burgess, *The Woodward Tornado of April 9, 1947* (Norman, OK: National Severe Storm Laboratory, 1988).

8 Irwin Walker ran: Dreyer, *The 1947 Woodward Oklahoma Tornado*.

8 After Ramona Kolander regained: Author interview with Kolander Esphahanian; Dreyer, *The 1947 Woodward Oklahoma Tornado*.

9 Shattuck Hospital's first indication: "Storm Kills 29 in West Texas, Hundreds Hurt," *The Daily Oklahoman*, Apr. 10, 1947, p. 1.

9 When Jim and Reva emerged: Dreyer, *The 1947 Woodward Oklahoma Tornado.*

9 The Stewart family: Ibid.

10 An enterprising telephone lineman: *The Dallas Morning News,* Apr. 11, 1947, p. 1.

10 A transport from Tinker: Bill Van Dyke, "Tinker Planes Bring in Storm Survivors," *The Daily Oklahoman,* Apr. 11, 1947, p. 3.

11 Still covered with dirt and blood: Author interview with Kolander Esphahanian; Dreyer, *The 1947 Woodward Oklahoma Tornado.*

11 "Sooner Hiroshima": Roy P. Stewart, "Sooner Hiroshima: The Big Wind—and Six," *The Daily Oklahoman,* Apr. 13, 1947, p. 1.

11 The man destined: T. T. Fujita, *Memoirs of an Effort to Unlock the Mystery of Severe Storms During the Fifty Years, 1942–1992* (Chicago: Wind Research Laboratory, Department of Geophysical Sciences, University of Chicago, 1992). All information about the life and scientific work of Dr. Fujita throughout this book, unless otherwise noted, comes from his self-published memoirs.

14 "We have spent two billion dollars": White House, "Statement by the President Announcing the Use of the A-Bomb at Hiroshima," Truman Presidential Museum and Library, Aug. 6, 1947, www.trumanlibrary.org/calendar/viewpapers.php?pid=100.

14 *Bockscar:* David Rezelman, et al., *The Manhattan Project: An Interactive History* (Department of Energy, Office of History and Heritage Resources, 2003); *The Manhattan Project: An Interactive History,* www.cfo.doe.gov/me70/manhattan/nagasaki.htm.

19 Saturday, three days after the tornado: Paul Swain, "Woodward Holds First Funeral Rites as Rain Adds to Misery," *The Daily Oklahoman,* Apr. 13, 1947, p. 1.

19 Tractors had to pull: "Storm Victims Leave Higgins," *The Dallas Morning News,* Apr. 13, 1947, p. 1.

19 Officially the toll stood: L. V. Wolford, "Preliminary Report on Tornadoes in the United States During 1947," in *United States Meteorological Yearbook for 1947* (Washington, D.C.: U.S. Weather Bureau, Dec. 1947), p. 247; Burgess, *The Woodward Tornado of April 9, 1947.* The initial tally was

167 dead with 95 fatalities in Woodward. More recent reassessments put the death toll at 181, with 107 dead in Woodward.

19 Tornadoes were a fact of life: National Weather Service, "The Highest Fatality-Producing Tornadoes in Oklahoma (1882–1998)," in *Fact Sheet for Tornadoes in the NWS Norman, Oklahoma, Area*, May 3, 1999.

20 As a boy: Gary A. England, *Weathering the Storm: Tornadoes, Television, and Turmoil* (Norman, OK: University of Oklahoma Press, 1996), pp. 3–4.

20 The 1947 tornado ranked sixth: Storm Prediction Center, "The 25 Deadliest U.S. Tornadoes," Historical Tornado Data Archive, www.spc.noaa.gov/archive/tornadoes/index.html.

20 From 1900 through 1940: Thomas P. Grazulis, *Significant Tornadoes 1890–1991* (St. Johnsbury, VT: Vermont Environmental Films, 1995); Marlene Bradford, *Scanning the Skies: A History of Tornado Forecasting* (Norman, OK: University of Oklahoma Press, 2001), p. 150.

20 But 1947 was an unusually deadly year: Wolford, *United States Meteorological Yearbook for 1947*, p. 247.

CHAPTER 2: A METEOROLOGICAL STAR

22 "We should have stayed": Author interviews with Roger Edwards and Richard Thompson, May 2004; conversational details recalled by Edwards in e-mail to the author, Feb. 2005.

23 The SPC was the nation's: National Oceanic and Atmospheric Administration, "Overview, About SPC," www.spc.noaa.gov/misc/aboutus.html.

28 And, in the end, it sometimes comes down to a hunch: Author interview with Edwards.

28 Nowhere else but America's midsection: Author interviews with Edwards and Thompson of the SPC, David Andra of the National Weather Service, and Harold Brooks of the National Severe Storms Laboratory, May 2004.

28 A tornado is the offspring: Edwards, "The Online Tornado FAQs."

28 The atmosphere seeks a constant: Author interview with Andra.

29 There are about 100,000 thunderstorms: National Weather Service, *Thunderstorms and Lightning . . . The Underrated Killers: A Preparedness Guide* (Washington, D.C.: National Weather Service, Jan. 1994), p. 2.

NOTES

29 On the Great Plains: Author interviews with Edwards, Thompson, and Jed Castles (meteorologist with KWTV Channel 9). A number of other books and articles helped inform the descriptions, including Jack William, *The Weather Book: An Easy-to-Understand Guide to the USA's Weather* (New York: Random House, 1997); David M. Ludlum, *National Audubon Society Field Guide to North American Weather* (New York: Alfred A. Knopf, 1991); John D. Cox, *Weather for Dummies* (New York: Hungry Minds, 2000); and "There's No Place Like Home: The Weather Issue," the May–June 2005 issue of *Oklahoma Today.*

30 Tornadoes are a peculiarly American problem: It is estimated that another 200 to 300 tornadoes occur annually in nations other than the United States. But most other nations lack the expansive radar, detection procedures, and methodical counting system operated by NWS.

30 But the first tornado ever recorded: Jack Williams, "Ask Jack," *USA Today,* May 30, 1997, www.usatoday.com/weather/resources/askjack/watorhty.htm.

31 Essentially a tornado is: Edwards, "The Online Tornado FAQs."

31 An average-size tornado: "A New Sort of Wind Power," *The Economist,* Sept. 29, 2005.

32 The whole chase scene: Author interview with Erik Rasmussen, June 2005.

33 It was the largest armada: "Frequently Asked Questions About Project VORTEX," NOAA news release, no. 95-48, July 11, 1995.

33 The odds of any given square mile: Federal Emergency Management Agency, "Midwest Tornadoes of May 3, 1999," in *Building Performance Assessment Report,* FEMA 342 (Washington, D.C.: U.S. Government Printing Office, Oct. 1999), pp. 2–4.

33 VORTEX-99: Keli Tarp, "Early Warnings Saved Lives on the Plains," *NOAA Report 8,* no. 6 (June 1999): 1–3.

33 The best odds of seeing: Author interview with Brooks.

34 The local NWS counted: Michael L. Branick, *Tornadoes in the Oklahoma City, Oklahoma, Area, Since 1890* (Norman, OK: NOAA Technical Memorandum NWS SR-160, 1994).

34 The first recorded tornado occurred in 1893: Ibid.

35 As the city recovered the 168 dead: England, *Weathering the Storm*, p. 206.

35 But in October 1998: Daniel McCarthy and Joseph T. Schaefer, *1998 Weather: Tornadoes* (Norman, OK: SPC publication, Jan. 1999).

35 Gary England: Author interview with England, May 2004.

36 Kara Wiese didn't know: Author interviews with Kara's mother, Mary; brother, Dustin; sister-in-law, Misty; and son, Jordan, Jan. 2005.

36 All she had in the world: U.S. Bankruptcy Court filing, United States Bankruptcy Court, Western District of Oklahoma, Oklahoma City, Case Number 99-11963, March 8, 1999, discharged June 24, 1999.

37 The only semblance of authority: Author interviews with Bridge Creek School superintendent Terry Brown and school secretary Janie Pruett, Jan. 2005.

38 Gary England's weather reports: Author interview with Edwards and Thompson; conversational details recalled by Edwards in e-mail to author, Feb. 2005.

CHAPTER 3: A TORNADO FORECAST

40 On March 20, 1948: Robert C. Miller and Charlie A. Crisp, "The Unfriendly Sky" (n.d.), unpublished manuscript. Unless otherwise noted, Miller's unpublished manuscript is the source for events surrounding the first tornado forecast.

42 Colleagues who worked with him: Robert Maddox, "Working with Col. M," www.stormtrack.org/library/archives/stmar99.htm, March–April 1999.

45 They also worked with William Maughan: Bradford, *Scanning the Skies*, p. 69.

46 The Tinker Duo reopened: Ibid., p. 71.

46 The Weather Bureau's roots: "NOAA's National Weather Service Celebrates 135th Anniversary," NOAA news release, Feb. 9, 2005.

46 In 1869, Abbe: W. J. Humphreys, "Dr. Cleveland Abbe," *Profiles in Time: Giants of Science*, NOAA History online, www.history.noaa.gov/giants/abbe .html.

46 Abbe urged the War Department: Cleveland Abbe, "Meteorology, 1859 to 1909, It Is Now a Science," *The New York Times*, Dec. 12, 1909,

NOTES

p. SMA3. Abbe penned a lengthy opinion page article extolling the advances in meteorology in the previous fifty years. The quotations attributed to Abbe in subsequent paragraphs come from this article.

46 "From the start": Humphreys, *Profiles in Time: Giants of Science.*

47 In 1877, a meticulous army enlistee: Joseph G. Galway, "J. P. Finley: The First Severe Storms Forecaster," *Bulletin of the American Meteorological Society* 66, no. 11 (Nov. 1985): 1389–1395.

47 He also served as a damage surveyor: Ibid.; Tim Marshall, "John Finley's First Tornado Damage Survey," *Stormtrack Newsletter* 23, no. 6 (Sept.–Oct. 2000). Marshall's article is based on Finley's report: *Tornadoes of May 29th and 30th, 1879, in Kansas, Missouri, Nebraska, and Iowa* (Washington, D.C.: Report of the Chief Signal Office, U.S. Signal Corps, 1880).

47 "The roaring of the storm": Finley, *Tornadoes of May 29th and 30th, 1879, in Kansas, Missouri, Nebraska, and Iowa.*

49 Finley became convinced: Galway, "J. P. Finley."

49 Overwhelmed by his numerous duties: Galway, "J. P. Finley."

50 The forecasts were short-lived: Galway, "J. P. Finley."

50 General Adolphous Greely: Bradford, *Scanning the Skies,* p. 44.

50 Finley was detailed briefly: Galway, "J. P. Finley."

51 "Formerly the instruction in meteorology": Abbe, "Meteorology, 1859 to 1909."

51 It would take World War I: Bradford, *Scanning the Skies,* pp. 20–21.

52 On March 18, 1925: "Interesting Quotes," *NOAA, NWS Tri-State Tornado* (Paducah, Ky.: NWS), www.crh.noaa.gov/pah/1925/iq_body.php.

52 Bradford noted: Bradford, *Scanning the Skies,* p. 55.

53 During World War II, the Weather Bureau did muster: Charles A. Doswell III, Alan R. Moller, and Harold E. Brooks, "Storm Spotting and Public Awareness Since the First Tornado Forecast of 1948," *Weather and Forecasting* 14, no. 4 (1999): 544–557.

53 The 1945 Thunderstorm Project: Roscoe R. Braham, Jr., "The Thunderstorm Project," speech to the Eighteenth Conference on Severe Local Storms Luncheon, *Bulletin of the American Meteorological Society* 77, no. 8 (Aug.

1996). Subsequent descriptions are based on this presentation.

54 Francis Reichelderfer: Bradford, *Scanning the Skies*, p. 78.

55 Tinker officials sent a news release: Henry W. "Wally" Kinnan, "Letter to the Editor," *Air Weather Association Newsletter*, 1998. Kinnan, formerly the public information officer at Tinker Air Force Base, issued the news release that forced the American Meteorological Society to open the Miller and Fawbush presentation to the public; www.airweaassn.org/archives/firsttndo.htm.

55 The Weather Bureau's Washington managers: Bradford, *Scanning the Skies*, p. 79.

55 In Oklahoma, the media: Ibid., pp. 79–80.

56 The Weather Bureau's forecasters: Ibid., pp. 85–86.

56 Reichelderfer chose five men: Ibid., p. 89. The five men were Joseph Galway, James Carr, Robert Martin, Allen Brunstein, and David Stowell, all chosen because they were young, college educated, and had received meteorology training in the military. Galway served not only as a pioneer for the predecessors of the Storm Prediction Center but developed important analytical tools, such as the lift index, to help predict storms. He retired after thirty-eight years but emerged as an important historical resource for early storm forecasting and as a biographer of John P. Finley. Like Fujita, Miller, and Byers, Galway also died in 1998.

57 The stogie-chewing Miller: Maddox, "Working with Col. M."

57 The year was one of the deadliest: SPC, "Tornado Deaths by Year and Month—1950–1999," *Historical Tornado Data Archive*, May 1999.

57 "If there were hydrogen bombs": International News Service, May 11, 1958. The impact of hydrogen bomb testing on the climate was a source of congressional concerns throughout the mid-1950s. Immediately after the 1953 tornadoes, the U.S. Air Force reassured Congress that the atomic testing program was not the cause of the violent storms. Congress rejected a proposal for a more thorough investigation of weapons testing on the weather.

CHAPTER 4: SEARCHING FOR CLUES
59 At midnight, Roger Edwards: Author interview with Edwards.

NOTES

59 Three forecasters were the minimum: NOAA, "Overview," *About the SPC*, www.spc.noaa.gov/misc/aboutus.html.

61 The big supercomputer: Roger Edwards, Stephen F. Corfidi, Richard L. Thompson, Jeffrey S. Evans, Jeffrey P. Craven, Jonathan P. Racy, and Daniel W. McCarthy of the SPC and Michael D. Vescio of the NWS Fort Worth Office, *Storm Prediction Center Forecasting Issues Related to the 3 May 1999 Tornado Outbreak* (Norman, OK: SPC, 2002).

62 Another unknown was lift: Author interview with Edwards and Thompson; Thompson and Edwards, "An Overview of Environmental Conditions and Forecast Implications of the 3 May 1999 Tornado Outbreak," *Weather and Forecasting* 15, no. 6 (2000): 682–699, available at www.spc.noaa.gov/publications/thompson/3may99/waf.htm.

62 Attached to the balloon was a radiosonde: "Remote Sensing with Radiosondes," *Jetstream, An Online Weather School*, www.srh.weather.gov/srh/jetstream/remote/ua.htm.

63 At 8:00 a.m., Roger issued: Author interview with Edwards and Thompson.

63 An SPC weather watch: Chris Hayes Novy of Southern Illinois University and Roger Edwards, David Imy, and Stephen Goss of SPC, *SPC and Its Products*, www.spc.noaa.gov/misc/about.html.

64 Before dawn: Author interview with the Wiese family.

64 The morning paper called for: *The Daily Oklahoman*, May 3, 1999, p. 8.

64 Kara arrived at work as early: Author interview with the Wiese family.

64 Kara and Jordan had spent: Ibid.

66 In the parking lot: Author interview with Craig Wolter, March 2005.

67 At 11:30 a.m., the Storm Prediction Center elevated: NOAA NWS, "Service Assessment, Oklahoma/Southern Kansas Tornado Outbreak of May 3, 1999," Aug. 1999, p. 2.

67 Wolter and his students: Author interview with Wolter.

68 The winds aloft blew: Junji Katsura and Harold Connor, "Destruction Mechanism of Oklahoma Tornadoes in 1999," *Journal of Natural Disaster Science* 24, no. 2 (2002): 61.

68 The midday sounding: Author interview with Edwards and Thompson.

68 A sheet of clouds: Ibid.

69 Josh's revolutionary idea: Author interview with Josh Wurman, June 2004.

69 Stein was giving the students: Author interview with Wolter: conversational details recalled by Wolter.

CHAPTER 5: THE TORNADO DETECTIVE

71 On foot and leading: Fujita, *Memoirs of an Effort to Unlock the Mystery of Severe Storms During the Fifty Years, 1942–1992.*

74 Before Fujita, tornado research: Gregory S. Forbes and Howard B. Bluestein, "Tornadoes, Tornadic Thunderstorms and Photogrammetry: A Review of the Contributions by T. T. Fujita," *Bulletin of the American Meteorological Society* 82, no. 1 (Jan. 2001): 74.

75 By chance, researchers with the Illinois State Water Survey: "The First Tornado Hook Echo Weather Radar Observations, April 9, 1953," in Illinois State Water Survey, *Achievements,* 2004, www.sws.uiuc.edu/hilites/achievements.asp.

75 Fujita's observational method of research: Forbes and Bluestein, "Tornadoes, Tornadic Thunderstorms and Photogrammetry," p. 89.

76 After trial and error, the United States responded: "Television Infrared Observation Satellite-1," *Destination Earth: 40+ Years of Earth Science,* National Aeronautics and Space Administration, www.earth.nasa.gov/history/tiros/tiros1.html; "Remote Sensing: NASA Remote Sensing Accomplishments," Earth Observatory Library, National Aeronautics and Space Administration, http://earthobservatory.nasa.gov/Library/RemoteSensing/remote_09.html.

76 It was Fujita who put the satellite clouds in motion: W. Paul Menzel, "Cloud Tracking with Satellite Imagery: From the Pioneering Work of Ted Fujita to the Present," *Bulletin of the American Meteorological Society* 82, no. 1 (Jan. 2001): 33–36.

77 "Ted was an idea man": Author interview with Charles A. Doswell III, June 2004.

77 Closer to the ground: Susan Cobb, "History of Weather Radar: Weather Radar Development Highlight of National Severe Storms Laboratory's First 40 Years," *NOAA Magazine,* story 151, Oct. 29, 2004.

77 The U.S. Weather Bureau created the National Severe: Ibid.

78 Palm Sunday 1965: William R. Deedler, "Palm Sunday Tornado Outbreak April 11th, 1965" (Mar. 2005), www.crh.noaa.gov/dtx/palmsunday/. Deedler was the weather historian for NWSFO Detroit/Pontiac, Michigan.

78 The Weather Bureau terminology: "NOAA Remembers the Midwest's Deadly 1965 Palm Sunday Tornado Outbreak," *NOAA Magazine,* story 2418, Apr. 11, 2005. Official details of the outbreak are from NOAA.

80 Five young engineering professors: "Wind Science and Engineering Center 35th Anniversary," Texas Tech University news release, May 10, 2005.

80 Fujita charted: Ibid.

80 "He was brilliant and meticulous": Author interview with McDonald, Apr. 2005.

82 The database provided researchers with: Forbes and Bluestein, "Tornadoes, Tornadic Thunderstorms and Photogrammetry," p. 75.

CHAPTER 6: PRIORITY ONE

83 At midday on May 3: Author interview with Gary England.

84 The 1947 Woodward tornado: England, *Weathering the Storm,* p. 5.

84 "On Sunday evenings, Harry": Author interview with England.

85 He also had a program on WKY radio: WKY Radio advertisement, *The Daily Oklahoman,* Sept. 22, 1952, p. 27.

85 Volkman, despite being wildly popular: "Classics Club Gets a Forecast," *Chicago Midweek,* May 5, 2004.

86 For a time, Oklahoma even considered: "Storm-Siren Planes Studied," *The Daily Oklahoman,* Apr. 9, 1952, p. 7.

86 A more efficient warning system: Kurt Pickering and Charles Bryant, *The History of Civil Defense and Emergency Management in Tennessee* (Nashville: Tennessee Emergency Management Agency, 2002), www.tnema.org/Archives/EMHistory/TNCDHistory1.htm. This report provides a social and historical perspective on civil defense and emergency management in general.

86 But the U.S. Weather Bureau: Roger C. Winton et al., "History of Operational Use of Weather Radar by U.S. Weather Service," *Weather and Forecasting* 13, no. 2 (1998): 219–43.

86 It was better than nothing: "Weatherman's Job Isn't All Sunshine," *The Daily Oklahoman*, June 10, 1956, p. 21.

87 Gary planned to attend college: England, *Weathering the Storm*, p. 21.

87 "There was no radar": Author interview with England.

88 In 1974, the Weather Bureau's: Maria Puente, "Early Warning System Saved Lives," *USA Today*, July 6, 1999.

88 "It was pretty obvious": Author interview with England.

88 In 1973, the National Severe Storm Laboratory's: Cobb, "History of Weather Radar."

89 It was, said Don Burgess: Author interview with Burgess, Apr. 2005.

89 That same Union City tornado: England, *Weathering the Storm*, pp. 69–71.

89 He convinced the Griffin family: England, *Weathering the Storm*, pp. 101–106.

90 "My rear end": Author interview with England.

90 "He didn't think we were": Author interview with Ken Crawford, Jan. 2005.

90 The low point in the relationship: England, *Weathering the Storm*, p. 179.

90 John V. Byrne, President Reagan's NOAA administrator: Philip J. Hilts, "U.S. Considers Selling Parts of Weather Service to Private Side," *The Washington Post*, Mar. 8, 1983, p. 1; Hilts, "Satellites for Sale: Reagan Set to Sell Weather Satellites, Wants to Cut Costs, Boost Business," *The Washington Post*, Mar. 9, 1983, p. 1.

91 Ten months after it was proposed: Philip J. Hilts, "Reagan Kills Satellite Sale," *The Washington Post*, Nov. 29, 1983, p. 1. Congress forced Reagan to kill the sale by amending an appropriations bill. Reagan signed the spending bill rather than continue the fight over privatization.

91 The National Weather Service entered the 1980s: *Weather Forecasting: Cost Growth and Delays in Billion-Dollar Weather Service Modernization*, GAO Report to Congress/IMTEC-92-12FS (Washington, D.C.: Government Accounting Office, Dec. 1991).

92 In 1979, it completed its formal report: Author interview with Burgess.

NOTES

92 Unisys, later bought out by Lockheed: "Weather Service Cites Radar Flaws," *The New York Times*, Mar. 24, 1991, p. 27.

92 In 1988, *The Washington Post* reported: Boyce Rensberger, "Forecast at U.S. Weather Service: Shortage of Funds for Modernization," *The Washington Post*, Mar. 14, 1988, p. 13.

92 In 1991, *Time*: Philip Elmer-DeWitt, "Why Forecasts Are Getting Cloudier," *Time*, July 1, 1991, p. 58.

93 Researchers later found: Kevin M. Simmons and Daniel Sutter, *WSR-88D: Tornado Warnings and Tornado Casualties*, NOAA Library, Aug. 24, 2004, p. 20.

94 By 1984, Tulsa was the most flood-prone city: Julia Johnson, "Turning the Tide in Flood Control," *American City and County Magazine*, Dec. 1, 1998; *Planning for a Sustainable Future: The Link Between Hazard Mitigation and Livability*, FEMA publication 364 (Washington, D.C.: Government Printing Office, Sept. 2000), p. 14; "FEMA Honors Tulsa as the Nation's Leading Floodplain Management Community," FEMA press release, Sept. 13, 2000.

94 Crawford envisioned a network: Author interview with Crawford.

96 In one case, the Oklahoma Mesonet even solved a murder: Dale A. Morris, Kevin A. Kloesel, and Kenneth C. Crawford, *OK-FIRST: A Six-Year Retrospective* (Norman, OK: Oklahoma Climatological Survey, 2002).

96 "There were no guarantees": Author interview with Crawford.

97 Enterprise Electronics: England, *Weathering the Storm*, pp. 188–189.

97 In an era of megamedia companies: Author interview with Joyce Reed, vice president, Griffin Communications, Jan. 2005.

97 In the past, Gary used an Exacto knife: Author interview with England.

97 In 1991, Val Castor, a Tulsa native: Author interview with Castor, Jan. 2005.

98 "All through the years, we would say": Author interview with England.

98 Gary had his own name for it: England, *Weathering the Storm*, pp. 185–195.

100 "The sky was chaotic": Author interview with England.

100 At the Storm Prediction Center, Rich: Author interview with Thompson.

100 At 3:23 p.m. he wrote: "CONVERGENCE: Thompson," Mesoscale Discussion 0345, Storm Prediction Center, May 3, 1999.

100 At 3:45: Interview with Thompson.

CHAPTER 7: HIDING FROM THE BEAR

102 The meteorological clues: "The Widespread Tornado Outbreak of April 3–4, 1974," National Disaster Survey Report, 74-1 (Washington, D.C.: NOAA, 1974).

102 The previous year set a record: Ibid.

102 As the low-pressure system moved eastward: "Analysis and Reconstruction of the 1974 Tornado Super Outbreak," *Risk Management Solutions*, Apr. 2, 2004, p. 2, www.rms.com/Publications/1974SuperTornadoReport.pdf.

103 The total: 148 tornadoes in 13 states: "Weather Service Commemorates Nation's Worst Tornado Outbreak," NOAA news release, Mar. 31, 1999.

103 Ted Fujita and his teams: Fujita, *Memoirs of an Effort to Unlock the Mystery of Severe Storms During the Fifty Years, 1942–1992*.

103 *National Geographic* dubbed him: Walter Orr Roberts, "We're Doing Something About the Weather!," *National Geographic* 141, no. 4 (Apr. 1972): 547.

103–04 He never shared his private life: Author interview with Greg Forbes, April 2005; e-mail exchange with Kazuya Fujita, June 2005.

104 And he always signed "Ted Fujita, Mr. Tornado": Author interview with Tom Skilling, Chicago's WKY-TV meteorologist.

104 "The core of science is to 'see' ": Leslie R. Lemon, "The Father of Mesometeorology," recollections provided to Tim Marshall for *Tribute to Dr. Fujita*, Nov. 1999, www.stormtrack.org/library/people/fujita.htm.

105 This time he landed and walked: Forbes and Bluestein, "Tornadoes, Tornadic Thunderstorms and Photogrammetry," p. 77.

105 Flying back to Chicago: Ibid.

105 His graduate students: Author interview with Forbes.

106 Besides, Fujita paid: *Weather Classroom*, "Fujita: A Life in the Storm" (The Weather Channel, 2003), video.

NOTES

106 A young Tom Grazulis: Tom Grazulis, "Personal Recollections of Ted Fujita," recollections provided to Tim Marshall for *Tribute to Dr. Fujita.*

107 Not all of Fujita's ideas were brilliant: Author interview with Doswell.

107 "Even if I am wrong 50 percent": Gregory S. Forbes and Roger Wakimoto, "Tetsuya Theodore Fujita: Meteorological Detective and Illustrator," recollections provided to Tim Marshall for *Tribute to Dr. Fujita.*

107 Its meteorologists in thirteen states: NOAA, *Natural Disaster Survey Report 74-1,* 1947.

109 Texas Tech's Kiesling: Author interview with Kiesling, Apr. 2005.

109 Generally, construction codes: Ibid.

110 A Texas Tech University report explained: James R. McDonald et al., *A Recommendation for an Enhanced Fujita Scale* (Lubbock, TX: Wind Science and Engineering Center, Texas Tech University, and the U.S. Department of Commerce, June 2004, rev. Jan. 2006), pp. 6–7.

110 Properly anchored, mobile homes can: Manufactured Housing Association of Oklahoma, "Myths and Reality," www.mhao.org/myths.asp.

111 In hurricane-prone areas, mobile homes: Ibid.

111 So associated with tornado damage: Ibid.

111 "In hurricane-prone regions": Author interview with Kiesling.

111 There were numerous reports: Rick Bragg, "Storm over South Florida Building Codes," *The New York Times,* May 27, 1999, p. A14; Judy Stark, "New Building Code Brings Cost, Confusion," *St. Petersburg* (Fla.) *Times,* Aug. 19, 2002.

111 The state in 2002 enacted a uniform: Stark, "New Building Code Brings Cost, Confusion."

111 Hurricane Andrew, with winds in excess of 175 miles: Ed Rappaport, "Preliminary Report on Hurricane Andrew (16–28 August, 1992)" (Miami: NOAA National Hurricane Center, 1993), www.nhc.noaa.gov/1992andrew .html.

112 There were some estimates: Stark, "New Building Code Brings Cost, Confusion."

112 The new Florida code requires: Abby Goodnough, "Hurricane Charley: After Storm, a New Look at Stiffer Building Codes," *The New York Times,*

Aug. 20, 2004, p. 20; Michael Quint, "A Storm over Housing Codes," *The New York Times*, Dec. 1, 1995, p. D-1; Stark, "New Building Code Brings Cost, Confusion."

112 The National Weather Service marked: NOAA news release, Mar. 23, 1999.

112 The Weather Service issued: Ibid.

Chapter 8: Inside the Bear's Cage

113 A meteorological army: Author interviews with Josh Wurman, Erik Rasmussen, Craig Wolter, and Val Castor.

114 Val Castor and his fiancée: Author interview with Castor.

114 At 4:15 p.m., the National Weather Service: "Oklahoma/Southern Kansas Tornado Outbreak of May 3, 1999," NOAA NWS Service Assessment (Aug. 1999), p. 2.

116 "Conditions are pretty favorable": KWTV News 9, *May Third, Deadly Force* (Oklahoma City: Griffin Television, LLC, 1999), video. All dialogue exchanges between Gary England and Val Castor and England's on-air public warnings come from this KWTV video.

116 Off camera, News 9 was crackling: Author interview with KWTV meteorologist Jed Castles, Jan. 2005.

116 At 4:45 p.m. the Storm Prediction Center: NOAA NWS Service Assessment, p. 9.

117 Two minutes after the SPC watch: Ibid.

117 At 4:51 p.m. Storm A: "May 3 Tornadoes in NWS Norman County Warning Area," NOAA NWS Preliminary Report, May 4, 1999.

117 The Minnesota kids piled: Author interview with Wolter.

119 After tossing the football: Author interview with Wiese family.

119 A second twister appeared: "May 3 Tornadoes in NWS Norman County Warning Area."

120 In the DOW truck: Author interview with Wurman.

121 Using OK-FIRST: Author interview with Steve Chapman, Jan. 2005.

122 Minutes before the twister: Ibid.

NOTES

122 A wing from a damaged plane: Greg Stumpf, National Severe Storms Laboratory, and James Ladue, NWS, NOAA NWSFO, Norman, preliminary local report, Public Information Statement, May 6, 1999.

122 "I was somewhat": Author interview with Doswell.

122 "Why don't you guys": Author correspondence with Edwards.

123 Moore resident Charlie Cusack: Author interview with Cusack, Apr. 2005.

123 Katherine Walton and her: Mark A. Hutchison and Ron Jackson, "Mother Sacrifices Life for Son," *The Daily Oklahoman*, May 8, 1999, p. 97.

123 At tiny Bridge Creek, Chad Erwin: Author interview with Erwin, Jan. 2005.

123 Across the street from the: Author interview with Janie Pruett, Jan. 2005.

125 Forecasters at the National Weather Service in Norman: Author interview with David Andra.

126 Tall, lean, middle-aged: Author interview with Terry Brown.

128 "TORNADO EMERGENCY": NOAA NWSFO, Norman, severe weather statement, copy provided by Andra.

129 The hail came first: Author interview with Erwin.

130 Kara and her mother: Author interview with Wiese family.

132 Tom Tinneman heard: "Tornado," *48 Hours*, CBS-TV, July 27, 2000.

132 Nearby, nineteen-year-old Amy Crago: "Tornado Week," *Storm Stories*, produced by The Weather Channel, 2003.

132 Down the road, Deon and Samantha Darnell: Ellie Sutter and Steve Lackmeyer, "Tiny Infant, Grandmother Lost to Storm, Dashing Dad's Hopes," *The Daily Oklahoman*, May 7, 1999, p. 129; "Tornado."

132 Wesley Early, who the previous: Christy Watson, "Bridge Creek Man Thankful for Wooden Lifesaver," *The Daily Oklahoman*, May 5, 1999, p. 24.

133 "Someone is getting": Doswell's personal Web site, http://webserv.chat systems.com/~doswell/OKC_may3rd/OKC_thoughts2.html.

NOTES

133 "I love you," she shouted: Hutchison and Jackson, "Mother Sacrifices Life for Son," p. 97.

135 Julie Rakestraw crawled: Diana Baldwin, "Stories Born in Death's Brush," *The Daily Oklahoman*, May 9, 1999, p. 1; "Angry Skies," *Naked Science*, produced by Pioneer Film & Television, National Geographic Channel, 2004.

135 He took one look at her: "Angry Skies."

135 Charlie Cusack sat on his couch: Author interview with Cusack.

137 Jack Damrill, doing play-by-play: Author interview with Damrill, Jan. 2005.

138 The slimly built Hanson crawled: Daniel J. Miller et al., "Highway Overpasses as Tornado Shelters: Fallout from the 3 May 1999 Oklahoma/Kansas Violent Tornado Outbreak," slide presentation at the National Weather Association annual meeting, Biloxi, Miss., Oct. 1999.

139 Tram Thu Bui: Melissa Nelson, "Woman Still Missing, Family Clings to Hope," *The Daily Oklahoman*, May 12, 1999, p. 1.

139 The Moore City Council had gathered: Author interview with city manager Steve Eddy, Jan. 2005.

140 Mall managers evacuated people: John Rohde, "OU Recruit White Takes Shelter in Stock Room," *The Daily Oklahoman*, May 5, 1999. The article refers to Jason White, who later won a Heisman trophy as a University of Oklahoma quarterback.

140 Trucker Anthony Batagglia saw it: "Baby Born Since Tornado Gives Mother New Joy," *The Daily Oklahoman*, Aug. 14, 1999, p. 121.

141 Carolyn Stager, a lobbyist for the Oklahoma: Author interview with Stager, Jan. 2005.

142 Mike Pederson, manager of the Cracker Barrel restaurant: Bryan Painter, "Man Avoids Flying Truck in Restaurant," *The Daily Oklahoman*, May 4, 1999, p. 10.

143 Seminole County emergency manager: Author interview with Crawford.

143 "All the TV stations and radio stations": Author interview with Ben Springfield, Jan. 2005.

143 Wakefield, a strapping twenty-six-year-old: Author interview with Joey Wakefield, Jan. 2005.

144 Wurman and his crew had: Author interview with Wurman.

145 Roger and Rich had: Author interview with Edwards and Thompson.

145 Gary had only one storm: Author interview with England.

145 Roger and Rich pulled off: Author interview with Edwards and Thompson.

146 Logan County emergency manager John Lewis: Dale A. Morris et al., *OK-FIRST: A Six-Year Retrospective.*

146 B20 temporarily put: According to the *Guinness Book of World Records*, the diameter of a tornado is defined as the distance between the rotational wind speed maxima: www.guinnessworldrecords.com/content_p.s/record .asp?recordid=581.

146 By the time B20 set down: Author interview with Wiese family.

CHAPTER 9: MR. TORNADO SEES HIS FIRST

Fujita, *Memoirs of an Effort to Unlock the Mystery of Severe Storms During the Fifty Years, 1942–1992; The Weather Classroom*, "Fujita: A Life in the Storm" (The Weather Channel, 2003), video.

CHAPTER 10: A TWISTER'S JOURNEY

150 This is the first sight: Author interview with Wiese family.

150 "How's the house?": Author interview with Erwin.

150 The tornado entered Bridge Creek: NOAA National Climatic Data Center, "Event Record, May 3, 1999, Tornado A9."

151 Grady County deputy Robert Jolley: Christy Watson, "Deputy Finds Tiny Survivor of Bridge Creek Tornado," *The Daily Oklahoman*, May 8, 1999, p. 15.

152 The National Weather Service would: NOAA NCDC Event Record, NOAA National Climatic Data Center, "Event Record, May 3, 1999, Tornado A9."

152 By the time Terry Brown: Author interview with Brown.

NOTES

153 Paramedic Steve Finley arrived: *Paramedics: On the Edge,* produced by New York Times Television and Discovery Health Channel, 2001; "Tornado," *48 Hours,* CBS, July 27, 2000.

154 School secretary Janie Pruett: Author interview with Pruett.

154 "At one point, they": "Tornado"; *Paramedics: On the Edge.*

154 As soon as the phone: Author interview with Wiese family.

155 The intensity of A9 had decreased: NOAA NCDC Event Record, NOAA National Climatic Data Center, "Event Record, May 3, 1999, Tornado A9."

156 "It looked": Randy Ellis, "Tornadoes Shred State," *The Daily Oklahoman,* May 4, 1999, p. 1.

156 Oklahoma City firefighters spray-painted: Melissa Nelson, "Victim Anger Flaring," *The Daily Oklahoman,* May 5, 1999, p. 1; Randy Ellis and Steve Lackmeyer, "Death Toll Raising Questions," *The Daily Oklahoman,* May 6, 1999, p. 1.

156 John Graham, a six-foot-five: Author interview with Graham, Jan. 2005.

157 Stunned doctors at Hillcrest Hospital: "Angry Skies."

158 All the ordinary landmarks: Author interview with Cusack.

158 The tornado's intensity fluctuated dramatically: NOAA NCDC Event Record, NOAA National Climatic Data Center, "Event Record, May 3, 1999, Tornado A9."

159 In Del City, Carolyn Stager: Author interview with Stager.

159 One of the first reporters to the Del Air: Author interview with Damrill.

160 The Weather Service survey: NOAA NCDC Event Record, NOAA National Climatic Data Center, "Event Record, May 3, 1999, Tornado A9."

160 Rescuers picked: Bobby Ross, Jr., and Melissa Nelson, "Storm Damage Evokes Images of War Zone," *The Daily Oklahoman,* May 5, 1999, p. 1.

160 In Moore, the fire chief surveyed: Author interview with Moore City manager Steve Eddy.

160 As the tornado entered the: Author interview with Wolter.

NOTES

161 Roger and Rich watched: Author interview with Edwards and Thompson.

162 Bridge Creek school secretary Janie Pruett: Author interview with Pruett.

165 "The analogy with a war zone": Doswell's personal Web site, http:// webserv.chatsystems.com/~doswell/OKC_may3rd/OKC_thoughts2.html.

CHAPTER 11: VORTEX

166 Weather geeks piled: Jenn Shreve, "Storm Chaser," *Salon*, July 19, 1999.

166 On stormy days, he: Author interview with Bluestein, Apr. 2005.

167 Scientists at Los Alamos National Laboratory had an even: Howard Bluestein, "A History of Severe-Storm-Intercept Field Programs," *Weather and Forecasting* 14, no. 4 (1999): 558–577.

167 Another scientist tried: Ibid.

167 Howie and crew tried: Ibid.

168 In the late 1980s, OU researchers: Ibid.

168 Using the portable Doppler: National Science Foundation, "Tracking Tornadoes, Nature's Most Powerful Winds," *Frontiers*, Feb. 1997, www.nsf .gov/news/frontiers_archive/2-97/index.jsp.

168 Josh Wurman took a look at Bluestein's: Author interview with Wurman.

169 "I don't care what": Richard Witkin, "Tower Was Urged to Shift Runways Before Jet Crash," *The New York Times*, July 1, 1975, p. 1.

170 Fujita began his own analysis: Walter Sullivan, "New Study of Downdrafts Expected to Reduce Danger of Plane Crashes," *The New York Times*, July 20, 1982, p. C1; James W. Wilson and Roger M. Wakimoto, "The Discovery of the Downburst: T. T. Fujita's Contribution," *Bulletin of the American Meteorological Society* 82, no. 1 (Jan. 2001): 49–68.

170 "Such a damage-causing": Fujita, *Memoirs of an Effort to Unlock the Mystery of Severe Storms During the Fifty Years, 1942–1992*.

173 Fujita conducted his own study: "Wind Bursts Blamed for Eight Air Accidents," *The New York Times*, Jan. 17, 1984, p. C4.

NOTES

173 Said one meteorologist: Fujita, *Memoirs of an Effort to Unlock the Mystery of Severe Storms During the Fifty Years, 1942–1992*, p. 110.

174 *The New York Times* quoted one: Editorial, "Fear of Flying," *The New York Times*, Aug. 29, 1985, p. A22.

174 "However, the microburst": Wilson and Wakimoto, "The Discovery of the Downburst," p. 60.

176 "In VORTEX, we wanted to figure": Author interview with Rasmussen.

CHAPTER 12: THE TWISTER'S AFTERMATH

178 "I should have held on tighter": Author interview with Wiese family; Patty Reinert, "Sorrowful Search for Missing Mom," *Houston Chronicle*, May 7, 1999, p. 1.

180 And he still loved to chase: Author interview with Marshall, Apr. 2005.

180 "It's clear that the construction": Author interview with Doswell.

180 Marshall found one house: Author interview with Marshall.

182 "We found violations here": Ibid.

182 Five years earlier, the Manufactured: Harold Brooks and Jim Purpura, "Mobile Home Tornado Fatalities: Some Observations," *AWARE* (Mar. 1994), www.nssl.noaa.gov/users/brooks/public_html/essays/mobilehome .html.

183 Chad Erwin, Kara's neighbor, could: Author interview with Erwin.

183 The statistics were staggering: National Weather Service, *Oklahoma/ Southern Kansas Tornado Outbreak of May 3, 1999*, Service Assessment, NWS, NOAA (Silver Spring, MD: Aug. 1999), www.nws.noaa.gov/om/ assessments/ok-ks/report7.pdf.

184 Gary England was stunned by: Author interview with England.

186 Eddy would have plenty of offers: Author interview with Eddy.

188 Over the days and weeks after the tornado: Author interview with Cusack.

189 Carolyn Stager lived with: Author interview with Stager.

191 Their report: Roger Edwards et al., "Storm Prediction Center Forecast-

ing Issues Related to the 3 May 1999 Tornado Outbreak," *Weather and Forecasting* 17, no. 3 (2002): 544–558.

191 For all of 1999, Oklahoma had 145 tornadoes: "1999 Oklahoma Tornadoes Break Record," NWS news release, Dec. 1999.

192 It had worked with FEMA: FEMA, *Taking Shelter from the Storm: Building a Safe Room Inside Your House,* FEMA 320 (Washington, D.C.: FEMA, Mar. 2004).

192 President Clinton: White House transcript, "Remarks by the President after Viewing Tornado Damage," May 8, 1999.

192 Texas Tech's Ernst Kiesling: Author interview with Kiesling.

193 Doswell noted that the entire warning system: Author interview with Doswell.

193 The city of Moore: Author interview with Eddy.

193 At Bridge Creek, Terry Brown: Author interview with Brown.

194 Chad Erwin bought a new: Author interview with Erwin.

194 On a hot August day in 2001: Author interview with Cusack.

Chapter 13: Seeing the Winds

196 In March 1998: NSSL, "Golden Anniversary of Tornado Forecasting: Fifty Years of Service to the American People," www.nssl.noaa.gov/GoldenAnniversary/; Charlie A. Crisp, "Fiftieth Anniversary of the First Tornado Forecast to Be Celebrated March 23–25, 1998," NSSL Briefings, Vol. 2, no. 2, Winter 1998.

198 Brian Smith, one of Fujita's grad students: Smith, "Working with Dr. Fujita," recollections provided to Tim Marshall for *Tribute to Dr. Fujita,* Nov. 1999, www.stormtrack.org/library/people/fujita.htm.

198 Fujita readily accepted speaking: Author interview with Tom Skilling, Apr. 2005.

198 Greg Forbes, another of Fujita's grad students: Author interview with Forbes.

198 Fujita's son, Kazuya: Author e-mail exchange with K. Fujita.

199 Still, Forbes and others were surprised: Author interview with Forbes.

199 "Ted always said that": Roger M. Wakimoto, "Prologue," *Bulletin of the American Meteorological Society* 82, no. 1 (Jan. 2001): 9.

199 *Weatherwise* magazine called: Jeff Rosenfeld, "Mr. Tornado: The Life and Career of Ted Fujita," *Weatherwise,* May 1, 1999.

199 "I was thrilled to call Ted": Wakimoto, "Prologue," p. 9.

200 The university dismantled: Author interview with McDonald.

200 There would be a Buddhist funeral: Author interview with Tom Skilling.

200 "The computer," he said, "doesn't understand": "Tetsuya 'Ted' Fujita, 1920–1998," University of Chicago news release, Nov. 20, 1998.

CHAPTER 14: A TORNADO'S GRIP

201 The Meatwagon met a horrible end: Author correspondence with Edwards.

201 In its dry language to its forecasters: National Centers for Environmental Prediction, Special NCEP Discussion Central Operations, NCEP fact sheet, Sept. 30, 1999.

202 NOAA's computer can make 450 billion: "New Weather and Climate Supercomputer Helps Advance NOAA Weather Service Forecasts," *NOAA Magazine,* story 1156, June 6, 2003.

202 "I refuse to feel a shred": Edwards personal Web site, www.stormeyes .org/tornado/3may99/comments.htm.

203 The Oklahoma Mesonet: "Oklahoma Programs Named Semifinalists in Government Innovations Award Program," Innovations in Government news release, Apr. 24, 2000.

203 Ken Crawford was named by NOAA: "Program Director Named to Lead COOP Modernization, Crawford Will Develop National Integrated Surface Observation System," NOAA news release, June 21, 2004.

203 President Bill Clinton toured: "Clinton: I Have Never Seen So Much Destruction," CNN.com; White House transcript, "Remarks by the President after Viewing Tornado Damage," May 8, 1999.

203 The $67 million, seven-story facility: "NOAA and University of Okla-

homa Break Ground at New National Weather Center Site," *NOAA Magazine,* story 64, Nov. 15, 2002.

203 The National Weather Center: "History of the Weather Center," University of Oklahoma, www.nwc.ou.edu/history.php.

204 Rick Santorum, a U.S. senator from: Maeve Reston, "Storm over Weather Service Initiatives," *Pittsburgh Post-Gazette,* Apr. 26, 2005, p. A1.

204 A former radio DJ, Charlie: Author interview with Cusack.

205 Gary called his new: Author interview with England.

205 For several miles: "Information for the May 8, 2003, Oklahoma City Area Tornadoes," NOAA NWS, Norman.

206 They called it the Enhanced Fujita: James R. McDonald et al., *A Recommendation for an Enhanced Fujita Scale,* pp. 6–7.

207 "Mitigating future losses": "Midwest Tornadoes of May 3, 1999," FEMA Building Performance Assessment Report, no. 342 (Oct. 1999), p. 8-1.

207 Its conclusion: Ibid., p. 7-1.

207 On the afternoon of May 3: Author interview with Rasmussen.

207 "It is hard to gather the data": Ibid.

208 VORTEX2: VORTEX2 Steering Committee, *Verification of the Origins of Rotation in Tornadoes Experiment, Experimental Design Overview,* Jan. 31, 2005, www.eol.ucar.edu/dir_off/OFAP/info/longterm/edo_VORTEX2.pdf.

209 In Houston: Author interview with Kolander Esphahanian.

ABOUT THE AUTHOR

Nancy Mathis, a native of Oklahoma, was a veteran journalist and former White House correspondent before going into media relations. She lives in Silver Spring, Maryland.